CELESTIAL AUTHORS

THE
INHABITED
UNIVERSE

Selected Papers from
the Urantia Revelation

Compiled by Saskia Praamsma

SQUARE
CIRCLES
PUBLISHING

THE INHABITED UNIVERSE

By Celestial Authors

Compiled by Saskia Praamsma

from selected papers of the Urantia revelation,

first published in 1955 as *The Urantia Book*.

Cover: Syrp & Co.

Collage images from www.nNasa.gov/copyright Triff, 2014

Used under license from Shutterstock.com

ISBN:978-0-9905813-0-7

SQUARE CIRCLES PUBLISHING

Associated websites:

SquareCirclesPublishing.com

SquareCircles.com

LifeOnOtherWorlds.com

Contents

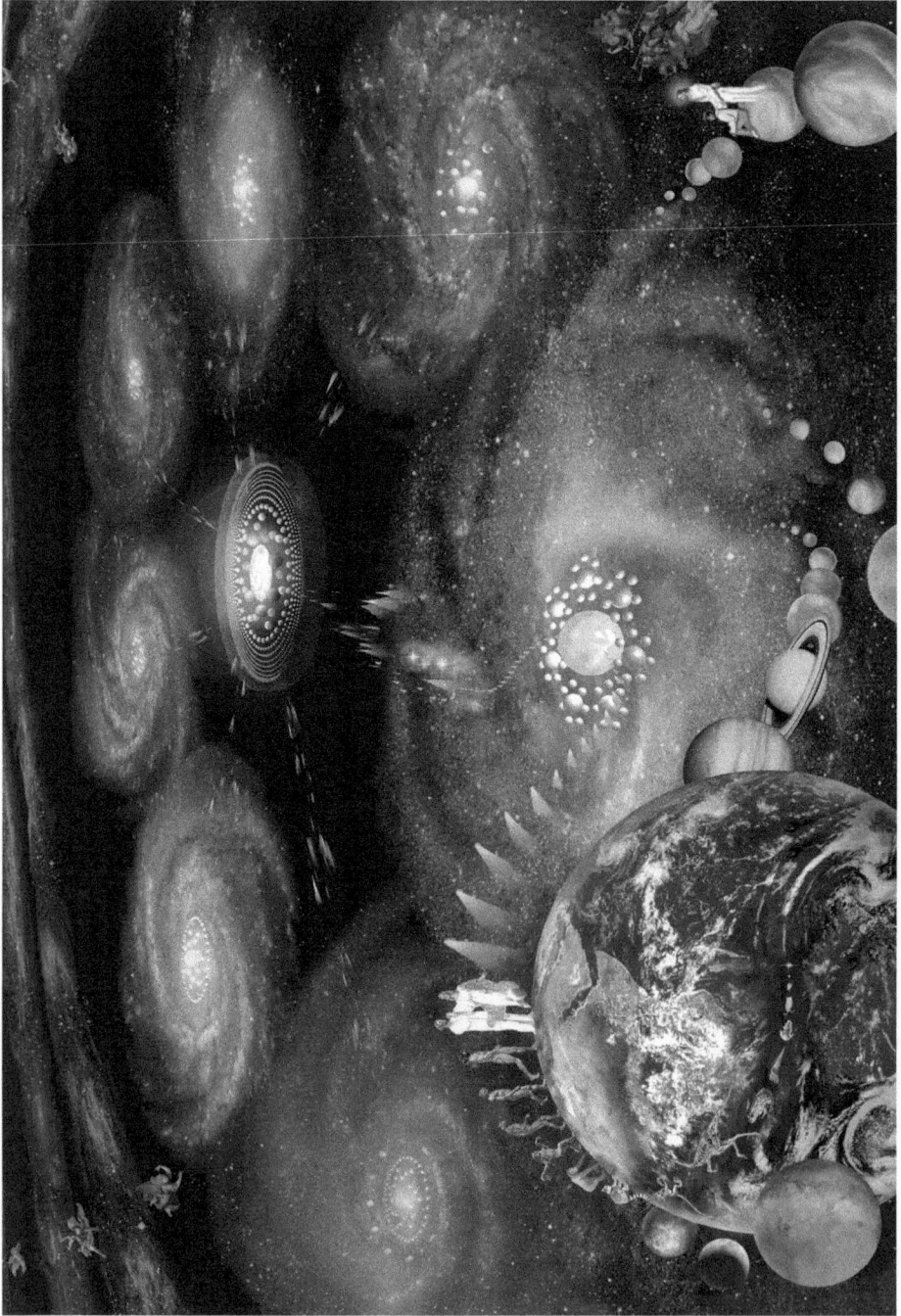

An artist's concept of the Master Universe, as portrayed in the Urantia revelation. [© Chick Montgomery and Saskia Praamsmal 1999]

INTRODUCTION

THE Urantia revelation was presented to our world by celestial beings through an anonymous human receiver. It was published in 1955 as *The Urantia Book*, a 2097-page volume divided into 196 papers. These papers describe God, the grand universe and its many inhabitants, our part of the cosmos, life after death, the origin and history of our planet, the life and teachings of Jesus, and the glorious destiny God has planned for every creature throughout his vast domains—an endless life of boundless opportunity and unlimited progress.

The following passages from the Foreword of the book, transmitted by a superhuman personality who calls himself "an Orvonton Divine Counselor, Chief of the Corps of Superuniverse Personalities assigned to portray on Urantia the truth concerning the Paradise Deities and the universe of universes," gives us a quick picture of the layout of the Grand Universe:

> Your world, Urantia, is one of many similar inhabited planets which comprise the local universe of Nebadon. This universe, together with similar creations, makes up the superuniverse of Orvonton, from whose capital, Uversa, our commission hails. Orvonton is one of the seven evolutionary superuniverses of time and space which circle the never-beginning, never-ending creation of divine perfection—the central universe of Havona. At the heart of this eternal and central universe is the stationary Isle of Paradise, the geographic center of infinity and the dwelling place of the eternal God.

The seven evolving superuniverses in association with the central and divine universe, we commonly refer to as the grand universe; these are the now organized and inhabited creations. They are all a part of the master universe, which also embraces the uninhabited but mobilizing universes of outer space.

This selection of papers focuses on the many worlds we will encounter as we journey through the universe, worlds teeming with life, both human and superhuman.

SASKIA PRAAMSMA
Compiler

OVERVIEW
FROM THE PAPERS

These papers do not—cannot—even begin to exhaust the story of the living creatures, creators, eventuators, and still-otherwise-existent beings who live and worship and serve in the swarming universes of time and in the central universe of eternity. (30:1.114[1])

* * *

Paradise is the geographic center of infinity (11:9.2). . . the center of all creation, the source of all energies, and the place of primal origin of all personalities. . . . [To] mortals the most important thing about eternal Paradise is the fact that this perfect abode of the Universal Father is the real and far-distant destiny of the immortal souls of the mortal and material sons of God, the ascending creatures of the evolutionary worlds of time and space. (11:9.8)

* * *

Gravity is the all-powerful grasp of the physical presence of Paradise . . . the omnipotent strand on which are strung the gleaming stars, blazing suns, and whirling spheres which constitute the universal physical adornment of the eternal God, who is all things, fills all things, and in whom all things consist. (11:8.1)

* * *

The Father indeed abides on Paradise, but his divine presence also dwells in the minds of men. (12:7.13)

---

[1] Refers to Paper:Section.Paragraph of *The Urantia Book*.

"God is spirit," but Paradise is not. The material universe is always the arena wherein take place all spiritual activities; spirit beings and spirit ascenders live and work on physical spheres of material reality. (12:8.1)

* * *

The universes of space and their component systems and worlds are all revolving spheres, moving along the endless circuits of the master universe space levels. Absolutely nothing is stationary in all the master universe except the very center of Havona, the eternal Isle of Paradise, the center of gravity. (12:4.1)

* * *

Spirit beings do not dwell in nebulous space; they do not inhabit ethereal worlds; they are domiciled on actual spheres of a material nature, worlds just as real as those on which mortals live. The Havona worlds are actual and literal, albeit their literal substance differs from the material organization of the planets of the seven superuniverses. (14:2.1)

* * *

Your planet is a member of an enormous cosmos; you belong to a well-nigh infinite family of worlds. . . (15:14.9) The grand universe number of your world, Urantia, is 5,342,482,337,666. That is the registry number on Uversa and on Paradise, your number in the catalogue of the inhabited worlds (15:14.8) . . . Urantia [is] number 606 in the planetary group, or system, of Satania. This system has at present 619 inhabited worlds, and more than two hundred additional planets are evolving favorably toward becoming inhabited worlds at some future time. (15:14.5)

* * *

Life does not originate spontaneously. Life is constructed according to plans formulated by the (unrevealed) Architects of Being and appears on the inhabited planets either by direct importation or as a result of the operations of the Life Carriers of the local universes. These carriers of life are among the most interesting and versatile of the diverse family of universe Sons. They are intrusted with designing and carrying creature life to the planetary spheres.

And after planting this life on such new worlds, they remain there for long periods to foster its development. (36:0.1)

* * *

Love of adventure, curiosity, and dread of monotony—these traits inherent in evolving human nature—were not put there just to aggravate and annoy you during your short sojourn on earth, but rather to suggest to you that death is only the beginning of an endless career of adventure, an everlasting life of anticipation, an eternal voyage of discovery. (14:5.10)

* * *

On the mansion worlds the resurrected mortal survivors resume their lives just where they left off when overtaken by death. When you go from Urantia to the first mansion world, you will notice considerable change, but if you had come from a more normal and progressive sphere of time, you would hardly notice the difference except for the fact that you were in possession of a different body; the tabernacle of flesh and blood has been left behind on the world of nativity. (47:3.1)

* * *

There is a great and glorious purpose in the march of the universes through space. All of your mortal struggling is not in vain. We are all part of an immense plan, a gigantic enterprise, and it is the vastness of the undertaking that renders it impossible to see very much of it at any one time and during any one life. We are all a part of an eternal project which the Gods are supervising and outworking. The whole marvelous and universal mechanism moves on majestically through space to the music of the meter of the infinite thought and the eternal purpose of the First Great Source and Center. (32:5.1)

ONE

THE ETERNAL ISLE
OF PARADISE

(PAPER 11)

PARADISE is the eternal center of the universe of universes and the abiding place of the Universal Father, the Eternal Son, the Infinite Spirit, and their divine co-ordinates and associates. This central Isle is the most gigantic organized body of cosmic reality in all the master universe. Paradise is a material sphere as well as a spiritual abode. All of the intelligent creation of the Universal Father is domiciled on material abodes; hence must the absolute controlling center also be material, literal. And again it should be reiterated that spirit things and spiritual beings are *real*.

The material beauty of Paradise consists in the magnificence of its physical perfection; the grandeur of the Isle of God is exhibited in the superb intellectual accomplishments and mind development of its inhabitants; the glory of the central Isle is shown forth in the infinite endowment of divine spirit personality—the light of life. But the depths of the spiritual beauty and the wonders of this magnificent ensemble are utterly beyond the comprehension of the finite mind of material creatures. The glory and spiritual splendor of the divine abode are impossible of mortal comprehension. And Paradise is from eternity; there are neither records nor traditions respecting the origin of this nuclear Isle of Light and Life.

1. The Divine Residence

Paradise serves many purposes in the administration of the universal realms, but to creature beings it exists primarily as the dwelling place of Deity. The personal presence of the Universal Father is resident at the very center of the upper surface of this well-nigh circular, but not spherical, abode of the Deities. This Paradise presence of the Universal Father is immediately surrounded by the personal presence of the Eternal Son, while they are both invested by the unspeakable glory of the Infinite Spirit.

God dwells, has dwelt, and everlastingly will dwell in this same central and eternal abode. We have always found him there and always will. The Universal Father is cosmically focalized, spiritually personalized, and geographically resident at this center of the universe of universes.

We all know the direct course to pursue to find the Universal Father. You are not able to comprehend much about the divine residence because of its remoteness from you and the immensity of the intervening space, but those who are able to comprehend the meaning of these enormous distances know God's location and residence just as certainly and literally as you know the location of New York, London, Rome, or Singapore, cities definitely and geographically located on Urantia. If you were an intelligent navigator, equipped with ship, maps, and compass, you could readily find these cities. Likewise, if you had the time and means of passage, were spiritually qualified, and had the necessary guidance, you could be piloted through universe upon universe and from circuit to circuit, ever journeying inward through the starry realms, until at last you would stand before the central shining of the spiritual glory of the Universal Father. Provided with all the necessities for the journey, it is just as possible to find the personal presence of God at the center of all things as to find distant cities on your own planet. That you have not visited these places in no way disproves their reality or actual existence. That so few of the universe creatures have found God on Paradise in no way disproves either the reality of his existence or the actuality of his spiritual person at the center of all things.

The Father is always to be found at this central location. Did he move, universal pandemonium would be precipitated, for there converge in him at this residential center the universal lines of gravity from the ends of creation. Whether we trace the personality circuit back through the universes or follow the ascending personalities as they journey inward to the Father; whether we trace the lines of material gravity to nether Paradise or follow the insurging cycles of cosmic force; whether we trace the lines of spiritual gravity to the Eternal Son or follow the inward processional of the Paradise Sons of God; whether we trace out the mind circuits or follow the trillions upon trillions of celestial beings who spring from the Infinite Spirit—by any of these observations or by all of them we are led directly back to the Father's presence, to his central abode. Here is God personally, literally, and actually present. And from his infinite being there flow the flood-streams of life, energy, and personality to all universes.

2. Nature of the Eternal Isle

Since you are beginning to glimpse the enormousness of the material universe discernible even from your astronomical location, your space position in the starry systems, it should become evident to you that such a tremendous material universe must have an adequate and worthy capital, a headquarters commensurate with the dignity and infinitude of the universal Ruler of all this vast and far-flung creation of material realms and living beings.

In form Paradise differs from the inhabited space bodies: it is not spherical. It is definitely ellipsoid, being one-sixth longer in the north-south diameter than in the east-west diameter. The central Isle is essentially flat, and the distance from the upper surface to the nether surface is one tenth that of the east-west diameter.

These differences in dimensions, taken in connection with its stationary status and the greater out-pressure of force-energy at the north end of the Isle, make it possible to establish absolute direction in the master universe.

The central Isle is geographically divided into three domains of activity:

1. Upper Paradise.

2. Peripheral Paradise.

3. Nether Paradise.

We speak of that surface of Paradise which is occupied with personality activities as the upper side, and the opposite surface as the nether side. The periphery of Paradise provides for activities that are not strictly personal or nonpersonal. The Trinity seems to dominate the personal or upper plane, the Unqualified Absolute the nether or impersonal plane. We hardly conceive of the Unqualified Absolute as a person, but we do think of the functional space presence of this Absolute as focalized on nether Paradise.

The eternal Isle is composed of a single form of materialization—stationary systems of reality. This literal substance of Paradise is a homogeneous organization of space potency not to be found elsewhere in all the wide universe of universes. It has received many names in different universes, and the Melchizedeks of Nebadon long since named it *absolutum*. This Paradise source material is neither dead nor alive; it is the original nonspiritual expression of the First Source and Center; it is *Paradise*, and Paradise is without duplicate.

It appears to us that the First Source and Center has concentrated all absolute potential for cosmic reality in Paradise as a part of his technique of self-liberation from infinity limitations, as a means of making possible subinfinite, even time-space, creation. But it does not follow that Paradise is time-space limited just because the universe of universes discloses these qualities. Paradise exists without time and has no location in space.

Roughly: space seemingly originates just below nether Paradise; time just above upper Paradise. Time, as you understand it, is not a feature of Paradise existence, though the citizens of the central Isle are fully conscious of nontime sequence of events. Motion is not inherent on Paradise; it is volitional. But the concept of distance, even absolute distance, has very much meaning as it may be applied to relative locations on Paradise. Paradise is nonspatial; hence its areas are absolute and therefore serviceable in many ways beyond the concept of mortal mind.

3. Upper Paradise

On upper Paradise there are three grand spheres of activity, the *Deity presence*, the *Most Holy Sphere*, and the *Holy Area*. The vast region immediately surrounding the presence of the Deities is set aside as the Most Holy Sphere and is reserved for the functions of worship, trinitization, and high spiritual attainment. There are no material structures nor purely intellectual creations in this zone; they could not exist there. It is useless for me to undertake to portray to the human mind the divine nature and the beauteous grandeur of the Most Holy Sphere of Paradise. This realm is wholly spiritual, and you are almost wholly material. A purely spiritual reality is, to a purely material being, apparently nonexistent.

While there are no physical materializations in the area of the Most Holy, there are abundant souvenirs of your material days in the Holy Land sectors and still more in the reminiscent historic areas of peripheral Paradise.

The Holy Area, the outlying or residential region, is divided into seven concentric zones. Paradise is sometimes called "the Father's House" since it is his eternal residence, and these seven zones are often designated "the Father's Paradise mansions." The inner or first zone is occupied by Paradise Citizens and the natives of Havona who may chance to be dwelling on Paradise. The next or second zone is the residential area of the natives of the seven superuniverses of time and space. This second zone is in part subdivided into seven immense divisions, the Paradise home of the spirit beings and ascendant creatures who hail from the universes of evolutionary progression. Each of these sectors is exclusively dedicated to the welfare and advancement of the personalities of a single superuniverse, but these facilities are almost infinitely beyond the requirements of the present seven superuniverses.

Each of the seven sectors of Paradise is subdivided into residential units suitable for the lodgment headquarters of one billion glorified individual working groups. One thousand of these units constitute a division. One hundred thousand divisions equal one congregation. Ten million congregations constitute an assembly. One billion assemblies make one grand unit. And this ascending series continues through the second grand unit, the third, and so on to the seventh grand unit. And seven of the grand units make up the master units, and seven of the master units constitute a superior unit; and thus by sevens the ascending series expands through the superior, supersuperior, celestial, supercelestial, to the supreme units. But even this does not utilize all the space available. This staggering number of residential designations on Paradise, a number beyond your concept, occupies considerably less than one per cent of the assigned area of the Holy Land. There is still plenty of room for those who are on their way inward, even for those who shall not start the Paradise climb until the times of the eternal future.

4. Peripheral Paradise

The central Isle ends abruptly at the periphery, but its size is so enormous that this terminal angle is relatively indiscernible within any circumscribed area. The peripheral surface of Paradise is occupied, in part, by the landing and dispatching fields for various groups of spirit personalities. Since the nonpervaded-space zones nearly impinge upon the periphery, all personality transports destined to Paradise land in these regions. Neither upper nor nether Paradise is approachable by transport supernaphim or other types of space traversers.

The Seven Master Spirits have their personal seats of power and authority on the seven spheres of the Spirit, which circle about Paradise in the space between the shining orbs of the Son and the inner circuit of the Havona worlds, but they maintain force-focal headquarters on the Paradise periphery. Here the slowly circulating presences of the Seven Supreme Power Directors indicate the location of the seven flash stations for certain Paradise energies going forth to the seven superuniverses.

Here on peripheral Paradise are the enormous historic and prophetic exhibit areas assigned to the Creator Sons, dedicated to the local universes of time and space. There are just seven trillion of these historic reservations now set up or in reserve, but these arrangements all together occupy only about four per cent of that portion of the peripheral area thus assigned. We infer that these vast reserves belong to creations sometime to be situated beyond the borders of the present known and inhabited seven superuniverses.

That portion of Paradise which has been designated for the use of the existing universes is occupied only from one to four per cent, while the area assigned to these activities is at least one million times that actually required for such purposes. Paradise is large enough to accommodate the activities of an almost infinite creation.

But a further attempt to visualize to you the glories of Paradise would be futile. You must wait, and ascend while you wait, for truly, "Eye has not seen, nor ear heard, neither has it entered into the mind of mortal man, the things which the Universal Father has prepared for those who survive the life in the flesh on the worlds of time and space."

5. NETHER PARADISE

Concerning nether Paradise, we know only that which is revealed; personalities do not sojourn there. It has nothing whatever to do with the affairs of spirit intelligences, nor does the Deity Absolute there function. We are informed that all physical-energy and cosmic-force circuits have their origin on nether Paradise, and that it is constituted as follows:

1. Directly underneath the location of the Trinity, in the central portion of nether Paradise, is the unknown and unrevealed Zone of Infinity.

2. This Zone is immediately surrounded by an unnamed area.

3. Occupying the outer margins of the under surface is a region having mainly to do with space potency and force-energy. The activities of this vast elliptical force center are not identifiable with the known functions of any triunity, but the primordial force-charge of space appears to be focalized in this area. This center consists of three concentric elliptical zones: The innermost is the focal point of the force-energy activities of Paradise itself; the outermost may possibly be identified with the functions of the Unqualified Absolute, but we are not certain concerning the space functions of the mid-zone.

The inner zone of this force center seems to act as a gigantic heart whose pulsations direct currents to the outermost borders of physical space. It directs and modifies force-energies but hardly drives them. The reality pressure-presence of this primal force is definitely greater at the north end of the Paradise center than in the southern regions; this is a uniformly registered difference. The mother force of space seems to flow in at the south and out at the north through the operation of some unknown circulatory system which is concerned with the diffusion of this basic form of force-energy. From time to time there are also noted differences in the east-west pressures. The forces emanating from this zone are not responsive to observable physical gravity but are always obedient to Paradise gravity.

The mid-zone of the force center immediately surrounds this area. This midzone appears to be static except that it expands and contracts through three cycles of activity. The least of these pulsations is in an east-west direction, the next in a north-south direction, while the greatest fluctuation is in every direction, a generalized expansion and contraction. The function of this mid-area has never been really identified, but it must have something to do with reciprocal adjustment between the inner and the outer zones of the force center. It is believed by many that the mid-zone is the control mechanism of the mid-space or quiet zones which separate the successive space levels of the master universe, but no evidence or revelation confirms this. This inference is derived from the knowledge that this mid-area is in some manner related to the functioning of the non-pervaded-space mechanism of the master universe.

The outer zone is the largest and most active of the three concentric and elliptical belts of unidentified space potential. This area is the site of unimagined activities, the central circuit point of emanations which proceed spaceward in every direction to the outermost borders of the seven superuniverses and on beyond to overspread the enormous and incomprehensible domains of all outer space. This space presence is entirely impersonal notwithstanding that in some undisclosed manner it seems to be indirectly responsive to the will and mandates of the infinite Deities when acting as the Trinity. This is believed to be the central focalization, the Paradise center, of the space presence of the Unqualified Absolute.

All forms of force and all phases of energy seem to be encircuited; they circulate throughout the universes and return by definite routes. But with the emanations of the activated zone of the Unqualified Absolute there appears to be either an outgoing or an incoming—never both simultaneously. This outer zone pulsates in agelong cycles of gigantic proportions. For a little more than one billion Urantia years the space-force of this center is outgoing; then for a similar length of time it will be incoming. And the space-force manifestations of this center are universal; they extend throughout all pervadable space.

All physical force, energy, and matter are one. All force-energy originally proceeded from nether Paradise and will eventually return thereto following the completion of its space circuit. But the energies and material organizations of the universe of universes did not all come from nether Paradise in their present phenomenal states; space is the womb of several forms of matter and prematter. Though the outer zone of the Paradise force center is the source of space-energies, space does not originate there. Space is not force, energy, or power. Nor do the pulsations of this zone account for the respiration of space, but the incoming and outgoing phases of this zone are synchronized with the two-billion-year expansion-contraction cycles of space.

6. SPACE RESPIRATION

We do not know the actual mechanism of space respiration; we merely observe that all space alternately contracts and expands. This respiration affects both the horizontal extension of pervaded space and the vertical extensions of unpervaded space which exist in the vast space reservoirs above and below Paradise. In attempting to imagine the volume outlines of these space reservoirs, you might think of an hourglass.

As the universes of the horizontal extension of pervaded space expand, the reservoirs of the vertical extension of unpervaded space contract and vice versa. There is a confluence of pervaded and unpervaded space just underneath nether Paradise. Both types of space there flow through the transmuting regulation channels, where changes are wrought making pervadable space nonpervadable and vice versa in the contraction and expansion cycles of the cosmos.

"Unpervaded" space means: unpervaded by those forces, energies, powers, and presences known to exist in pervaded space. We do not know whether vertical (reservoir) space is destined always to function as the equipoise of horizontal (universe) space; we do not know whether there is a creative intent concerning unpervaded space; we really know very little about the space reservoirs, merely that they exist, and that they seem to counterbalance the space-expansion-contraction cycles of the universe of universes.

The cycles of space respiration extend in each phase for a little more than one billion Urantia years. During one phase the universes expand; during the next they contract. Pervaded space is now approaching the mid-point of the expanding phase, while unpervaded space nears the mid-point of the contracting phase, and we are informed that the outermost limits of both space extensions are, theoretically, now approximately equidistant from Paradise. The unpervaded-space reservoirs now extend vertically above upper Paradise and below nether Paradise just as far as the pervaded space of the universe extends horizontally outward from peripheral Paradise to and even beyond the fourth outer space level.

For a billion years of Urantia time the space reservoirs contract while the master universe and the force activities of all horizontal space expand. It thus requires a little over two billion Urantia years to complete the entire expansioncontraction cycle.

7. SPACE FUNCTIONS OF PARADISE

Space does not exist on any of the surfaces of Paradise. If one "looked" directly up from the upper surface of Paradise, one would "see" nothing but unpervaded space going out or coming in, just now coming in. Space does not touch Paradise; only the quiescent *midspace zones* come in contact with the central Isle.

Paradise is the actually motionless nucleus of the relatively quiescent zones existing between pervaded and unpervaded space. Geographically these zones appear to be a relative extension of Paradise, but there probably is some motion in them. We know very little about them, but we observe that these zones of lessened space motion separate pervaded and unpervaded space. Similar zones once existed between the levels of pervaded space, but these are now less quiescent.

The vertical cross section of total space would slightly resemble a maltese cross, with the horizontal arms representing pervaded (universe) space and the vertical arms representing unpervaded (reservoir) space. The areas between the four arms would separate them somewhat as the midspace zones separate pervaded and unpervaded space. These quiescent midspace zones grow larger and larger at greater and greater distances from Paradise and eventually encompass the borders of all space and completely incapsulate both the space reservoirs and the entire horizontal extension of pervaded space.

Space is neither a subabsolute condition within, nor the presence of, the Unqualified Absolute, neither is it a function of the Ultimate. It is a bestowal of Paradise, and the space of the grand universe and that of all outer regions is believed to be actually pervaded by the ancestral space potency of the Unqualified Absolute. From near approach to peripheral Paradise, this pervaded space extends horizontally outward through the fourth space level and beyond the periphery of the master universe, but how far beyond we do not know.

If you imagine a finite, but inconceivably large, V-shaped plane situated at right angles to both the upper and lower surfaces of Paradise, with its point nearly tangent to peripheral Paradise, and then visualize this plane in elliptical revolution about Paradise, its revolution would roughly outline the volume of pervaded space.

There is an upper and a lower limit to horizontal space with reference to any given location in the universes. If one could move far enough at right angles to the plane of Orvonton, either up or down, eventually the upper or lower limit of pervaded space would be encountered. Within the known dimensions of the master universe these limits draw farther and farther apart at greater and greater distances from Paradise; space thickens, and it thickens somewhat faster than does the plane of creation, the universes.

The relatively quiet zones between the space levels, such as the one separating the seven superuniverses from the first outer space level, are enormous elliptical regions of quiescent space activities. These zones separate the vast galaxies which race around Paradise in orderly procession. You may visualize the first outer space level, where untold universes are now in process of formation, as a vast procession of galaxies swinging around Paradise, bounded above and below by the midspace zones of quiescence and bounded on the inner and outer margins by relatively quiet space zones.

A space level thus functions as an elliptical region of motion surrounded on all sides by relative motionlessness. Such relationships of motion and quiescence constitute a curved space path of lessened resistance to motion which is universally followed by cosmic force and emergent energy as they circle forever around the Isle of Paradise.

This alternate zoning of the master universe, in association with the alternate clockwise and counterclockwise flow of the galaxies, is a factor in the stabilization of physical gravity designed to prevent the accentuation of gravity pressure to the point of disruptive and dispersive activities. Such an arrangement exerts antigravity influence and acts as a brake upon otherwise dangerous velocities.

8. Paradise Gravity

The inescapable pull of gravity effectively grips all the worlds of all the universes of all space. Gravity is the all-powerful grasp of the physical presence of Paradise. Gravity is the omnipotent strand on which are strung the gleaming stars, blazing suns, and whirling spheres which constitute the universal physical adornment of the eternal God, who is all things, fills all things, and in whom all things consist.

The center and focal point of absolute material gravity is the Isle of Paradise, complemented by the dark gravity bodies encircling Havona and equilibrated by the upper and nether space reservoirs. All known emanations of nether Paradise invariably and unerringly respond to the central gravity pull operating upon the endless circuits of the elliptical space levels of the master universe. Every known form of cosmic reality has the bend of the ages, the trend of the circle, the swing of the great ellipse.

Space is nonresponsive to gravity, but it acts as an equilibrant on gravity. Without the space cushion, explosive action would jerk surrounding space bodies. Pervaded space also exerts an antigravity influence upon physical or linear gravity; space can actually neutralize such gravity action even though it cannot delay it. Absolute gravity is Paradise gravity. Local or linear gravity pertains to the electrical stage of energy or matter; it operates within the central, super-, and outer universes, wherever suitable materialization has taken place.

The numerous forms of cosmic force, physical energy, universe power, and various materializations disclose three general, though not perfectly clear-cut, stages of response to Paradise gravity:

1. *Pregravity Stages (Force)*. This is the first step in the individuation of space potency into the pre-energy forms of cosmic force. This state is analogous to the concept of the primordial force-charge of space, sometimes called *pure energy* or *segregata*.

2. *Gravity Stages (Energy)*. This modification of the force-charge of space is produced by the action of the Paradise force

organizers. It signalizes the appearance of energy systems responsive to the pull of Paradise gravity. This emergent energy is originally neutral but consequent upon further metamorphosis will exhibit the so-called negative and positive qualities. We designate these stages *ultimata*.

3. *Postgravity Stages (Universe Power)*. In this stage, energy-matter discloses response to the control of linear gravity. In the central universe these physical systems are threefold organizations known as *triata*. They are the superpower mother systems of the creations of time and space. The physical systems of the superuniverses are mobilized by the Universe Power Directors and their associates. These material organizations are dual in constitution and are known as *gravita*. The dark gravity bodies encircling Havona are neither triata nor gravita, and their drawing power discloses both forms of physical gravity, linear and absolute.

Space potency is not subject to the interactions of any form of gravitation. This primal endowment of Paradise is not an actual level of reality, but it is ancestral to all relative functional nonspirit realities—all manifestations of force-energy and the organization of power and matter. Space potency is a term difficult to define. It does not mean that which is ancestral to space; its meaning should convey the idea of the potencies and potentials existent within space. It may be roughly conceived to include all those absolute influences and potentials which emanate from Paradise and constitute the space presence of the Unqualified Absolute.

Paradise is the absolute source and the eternal focal point of all energy-matter in the universe of universes. The Unqualified Absolute is the revealer, regulator, and repository of that which has Paradise as its source and origin. The universal presence of the Unqualified Absolute seems to be equivalent to the concept of a potential infinity of gravity extension, an elastic tension of Paradise presence. This concept aids us in grasping the fact that everything is drawn inward towards Paradise. The illustration is crude but nonetheless helpful. It also explains why gravity always acts preferentially in the plane perpendicular to the mass, a phenomenon indicative of the differential dimensions of Paradise and the surrounding creations.

9. THE UNIQUENESS OF PARADISE

Paradise is unique in that it is the realm of primal origin and the final goal of destiny for all spirit personalities. Although it is true that not all of the lower spirit beings of the local universes are immediately destined to Paradise, Paradise still remains the goal of desire for all supermaterial personalities.

Paradise is the geographic center of infinity; it is not a part of universal creation, not even a real part of the eternal Havona universe. We commonly refer to the central Isle as belonging to the divine universe, but it really does not. Paradise is an eternal and exclusive existence.

In the eternity of the past, when the Universal Father gave infinite personality expression of his spirit self in the being of the Eternal Son, simultaneously he revealed the infinity potential of his nonpersonal self as Paradise. Nonpersonal and nonspiritual Paradise appears to have been the inevitable repercussion to the Father's will and act which eternalized the Original Son. Thus did the Father project reality in two actual phases—the personal and the nonpersonal, the spiritual and the nonspiritual. The tension between them, in the face of will to action by the Father and the Son, gave existence to the Conjoint Actor and the central universe of material worlds and spiritual beings.

When reality is differentiated into the personal and the nonpersonal (Eternal Son and Paradise), it is hardly proper to call that which is nonpersonal "Deity" unless somehow qualified. The energy and material repercussions of the acts of Deity could hardly be called Deity. Deity may cause much that is not Deity, and Paradise is not Deity; neither is it conscious as mortal man could ever possibly understand such a term.

Paradise is not ancestral to any being or living entity; it is not a creator. Personality and mind-spirit relationships are *transmissible*, but pattern is not. Patterns are never reflections; they are duplications—reproductions. Paradise is the absolute of patterns; Havona is an exhibit of these potentials in actuality.

God's residence is central and eternal, glorious and ideal. His home is the beauteous pattern for all universe headquarters worlds;

and the central universe of his immediate indwelling is the pattern for all universes in their ideals, organization, and ultimate destiny.

Paradise is the universal headquarters of all personality activities and the source-center of all force-space and energy manifestations. Everything which has been, now is, or is yet to be, has come, now comes, or will come forth from this central abiding place of the eternal Gods. Paradise is the center of all creation, the source of all energies, and the place of primal origin of all personalities.

After all, to mortals the most important thing about eternal Paradise is the fact that this perfect abode of the Universal Father is the real and far-distant destiny of the immortal souls of the mortal and material sons of God, the ascending creatures of the evolutionary worlds of time and space. Every God-knowing mortal who has espoused the career of doing the Father's will has already embarked upon the long, long Paradise trail of divinity pursuit and perfection attainment. And when such an animal-origin being does stand, as countless numbers now do, before the Gods on Paradise, having ascended from the lowly spheres of space, such an achievement represents the reality of a spiritual transformation bordering on the limits of supremacy.

[Presented by a Perfector of Wisdom commissioned thus to function by the Ancients of Days on Uversa.]

TWO

THE UNIVERSE
OF UNIVERSES

(PAPER 12)

THE immensity of the far-flung creation of the Universal Father is utterly beyond the grasp of finite imagination; the enormousness of the master universe staggers the concept of even my order of being. But the mortal mind can be taught much about the plan and arrangement of the universes; you can know something of their physical organization and marvelous administration; you may learn much about the various groups of intelligent beings who inhabit the seven superuniverses of time and the central universe of eternity.

In principle, that is, in eternal potential, we conceive of material creation as being infinite because the Universal Father is actually infinite, but as we study and observe the total material creation, we know that at any given moment in time it is limited, although to your finite minds it is comparatively limitless, virtually boundless.

We are convinced, from the study of physical law and from the observation of the starry realms, that the infinite Creator is not yet manifest in finality of cosmic expression, that much of the cosmic potential of the Infinite is still self-contained and unrevealed. To created beings the master universe might appear to be almost infinite, but it is far from finished; there are still physical limits to the material creation, and the experiential revelation of the eternal purpose is still in progress.

1. SPACE LEVELS OF THE MASTER UNIVERSE

The universe of universes is not an infinite plane, a boundless cube, nor a limitless circle; it certainly has dimensions. The laws of physical organization and administration prove conclusively that the whole vast aggregation of force-energy and matter-power functions ultimately as a space unit, as an organized and co-ordinated whole. The observable behavior of the material creation constitutes evidence of a physical universe of definite limits. The final proof of both a circular and delimited universe is afforded by the, to us, well-known fact that all forms of basic energy ever swing around the curved path of the space levels of the master universe in obedience to the incessant and absolute pull of Paradise gravity.

The successive space levels of the master universe constitute the major divisions of pervaded space—total creation, organized and partially inhabited or yet to be organized and inhabited. If the master universe were not a series of elliptical space levels of lessened resistance to motion, alternating with zones of relative quiescence, we conceive that some of the cosmic energies would be observed to shoot off on an infinite range, off on a straight-line path into trackless space; but we never find force, energy, or matter thus behaving; ever they whirl, always swinging onward in the tracks of the great space circuits.

Proceeding outward from Paradise through the horizontal extension of pervaded space, the master universe is existent in six concentric ellipses, the space levels encircling the central Isle:

1. The Central Universe—Havona.
2. The Seven Superuniverses.
3. The First Outer Space Level.
4. The Second Outer Space Level.
5. The Third Outer Space Level.
6. The Fourth and Outermost Space Level.

Havona, the central universe, is not a time creation; it is an eternal existence. This never-beginning, never-ending universe consists of one billion spheres of sublime perfection and is surrounded by the enormous dark gravity bodies. At the center of Havona is the stationary and absolutely stabilized Isle of Paradise, surrounded by its twenty-one satellites. Owing to the enormous encircling masses of the dark gravity bodies about the fringe of the central universe, the mass content of this central creation is far in excess of the total known mass of all seven sectors of the grand universe.

The Paradise-Havona System, the eternal universe encircling the eternal Isle, constitutes the perfect and eternal nucleus of the master universe; all seven of the superuniverses and all regions of outer space revolve in established orbits around the gigantic central aggregation of the Paradise satellites and the Havona spheres.

The Seven Superuniverses are not primary physical organizations; nowhere do their boundaries divide a nebular family, neither do they cross a local universe, a prime creative unit. Each superuniverse is simply a geographic space clustering of approximately one seventh of the organized and partially inhabited post-Havona creation, and each is about equal in the number of local universes embraced and in the space encompassed. *Nebadon*, your local universe, is one of the newer creations in *Orvonton*, the seventh superuniverse.

The Grand Universe is the present organized and inhabited creation. It consists of the seven superuniverses, with an aggregate evolutionary potential of around seven trillion inhabited planets, not to mention the eternal spheres of the central creation. But this tentative estimate takes no account of architectural administrative spheres, neither does it include the outlying groups of unorganized universes. The present ragged edge of the grand universe, its uneven and unfinished periphery, together with the tremendously unsettled condition of the whole astronomical plot, suggests to our star students that even the seven superuniverses are, as yet, uncompleted. As we move from within, from the divine center outward in any one direction, we do, eventually, come to the outer limits of the organized and inhabited creation; we come to the outer limits of the grand universe. And it is near this outer border, in a far-off corner of such a magnificent creation, that your local universe has its eventful existence.

The Outer Space Levels. Far out in space, at an enormous distance from the seven inhabited superuniverses, there are assembling vast and unbelievably stupendous circuits of force and materializing energies. Between the energy circuits of the seven superuniverses and this gigantic outer belt of force activity, there is a space zone of comparative quiet, which varies in width but averages about four hundred thousand light-years. These space zones are free from star dust—cosmic fog. Our students of these phenomena are in doubt as to the exact status of the space-forces existing in this zone of relative quiet which encircles the seven superuniverses. But about one-half million light-years beyond the periphery of the present grand universe we observe the beginnings of a zone of an unbelievable energy action which increases in volume and intensity for over twenty-five million light-years. These tremendous wheels of energizing forces are situated in the first outer space level, a continuous belt of cosmic activity encircling the whole of the known, organized, and inhabited creation.

Still greater activities are taking place beyond these regions, for the Uversa physicists have detected early evidence of force manifestations more than fifty million light-years beyond the outermost ranges of the phenomena in the first outer space level. These activities undoubtedly presage the organization of the material creations of the second outer space level of the master universe.

The central universe is the creation of eternity; the seven superuniverses are the creations of time; the four outer space levels are undoubtedly destined to eventuate-evolve the ultimacy of creation. And there are those who maintain that the Infinite can never attain full expression short of infinity; and therefore do they postulate an additional and unrevealed creation beyond the fourth and outermost space level, a possible ever-expanding, never-ending universe of infinity. In theory we do not know how to limit either the infinity of the Creator or the potential infinity of creation, but as it exists and is administered, we regard the master universe as having limitations, as being definitely delimited and bounded on its outer margins by open space.

2. THE DOMAINS
OF THE UNQUALIFIED ABSOLUTE

When Urantia astronomers peer through their increasingly powerful telescopes into the mysterious stretches of outer space and there behold the amazing evolution of almost countless physical universes, they should realize that they are gazing upon the mighty outworking of the unsearchable plans of the Architects of the Master Universe. True, we do possess evidences which are suggestive of the presence of certain Paradise personality influences here and there throughout the vast energy manifestations now characteristic of these outer regions, but from the larger viewpoint the space regions extending beyond the outer borders of the seven superuniverses are generally recognized as constituting the domains of the Unqualified Absolute.

Although the unaided human eye can see only two or three nebulae outside the borders of the superuniverse of Orvonton, your telescopes literally reveal millions upon millions of these physical universes in process of formation. Most of the starry realms visually exposed to the search of your present-day telescopes are in Orvonton, but with photographic technique the larger telescopes penetrate far beyond the borders of the grand universe into the domains of outer space, where untold universes are in process of organization. And there are yet other millions of universes beyond the range of your present instruments.

In the not-distant future, new telescopes will reveal to the wondering gaze of Urantian astronomers no less than 375 million new galaxies in the remote stretches of outer space. At the same time these more powerful telescopes will disclose that many island universes formerly believed to be in outer space are really a part of the galactic system of Orvonton. The seven superuniverses are still growing; the periphery of each is gradually expanding; new nebulae are constantly being stabilized and organized; and some of the nebulae which Urantian astronomers regard as extragalactic are actually on the fringe of Orvonton and are traveling along with us.

The Uversa star students observe that the grand universe is surrounded by the ancestors of a series of starry and planetary clusters which completely encircle the present inhabited creation as concentric rings of outer universes upon universes. The physicists of Uversa calculate that the energy and matter of these outer and uncharted regions already equal many times the total material mass and energy charge embraced in all seven superuniverses. We are informed that the metamorphosis of cosmic force in these outer space levels is a function of the Paradise force organizers. We also know that these forces are ancestral to those physical energies which at present activate the grand universe. The Orvonton power directors, however, have nothing to do with these far-distant realms, neither are the energy movements therein discernibly connected with the power circuits of the organized and inhabited creations.

We know very little of the significance of these tremendous phenomena of outer space. A greater creation of the future is in process of formation. We can observe its immensity, we can discern its extent and sense its majestic dimensions, but otherwise we know little more about these realms than do the astronomers of Urantia. As far as we know, no material beings on the order of humans, no angels or other spirit creatures, exist in this outer ring of nebulae, suns, and planets. This distant domain is beyond the jurisdiction and administration of the superuniverse governments.

Throughout Orvonton it is believed that a new type of creation is in process, an order of universes destined to become the scene of the future activities of the assembling Corps of the Finality; and if our conjectures are correct, then the endless future may hold for all of you the same enthralling spectacles that the endless past has held for your seniors and predecessors.

3. Universal Gravity

All forms of force-energy—material, mindal, or spiritual—are alike subject to those grasps, those universal presences, which we call gravity. Personality also is responsive to gravity—to the Father's exclusive circuit; but though this circuit is exclusive to the Father, he is not excluded from the other circuits; the Universal Father is infinite and acts over *all* four absolute-gravity circuits in the master universe:

1. The Personality Gravity of the Universal Father.
2. The Spirit Gravity of the Eternal Son.
3. The Mind Gravity of the Conjoint Actor.
4. The Cosmic Gravity of the Isle of Paradise.

These four circuits are not related to the nether Paradise force center; they are neither force, energy, nor power circuits. They are absolute *presence* circuits and like God are independent of time and space.

In this connection it is interesting to record certain observations made on Uversa during recent millenniums by the corps of gravity researchers. This expert group of workers has arrived at the following conclusions regarding the different gravity systems of the master universe:

1. *Physical Gravity.* Having formulated an estimate of the summation of the entire physical-gravity capacity of the grand universe, they have laboriously effected a comparison of this finding with the estimated total of absolute gravity presence now operative. These calculations indicate that the total gravity action on the grand universe is a very small part of the estimated gravity pull of Paradise, computed on the basis of the gravity response of basic physical units of universe matter. These investigators reach the amazing conclusion that the central universe and the surrounding seven superuniverses are at the present time making use of only about five per cent of the active functioning of the Paradise absolute-gravity grasp. In other words: At the present moment about ninety-five per cent of the active cosmic-gravity action of the Isle of Paradise, computed on this totality theory, is engaged in controlling material systems beyond the borders of the present organized universes. These calculations all refer to absolute gravity; linear gravity is an interactive phenomenon which can be computed only by knowing the actual Paradise gravity.

2. *Spiritual Gravity.* By the same technique of comparative estimation and calculation these researchers have explored the present reaction capacity of spirit gravity and, with the co-operation of Solitary Messengers and other spirit personalities, have arrived at the summation of the active spirit gravity of the Second Source and Center. And it is most instructive to note that they find about the same value for the actual and functional presence of spirit gravity in the grand universe that they postulate for the present total of active spirit gravity. In other words: At the present time practically the entire spirit gravity of the Eternal Son, computed on this theory of totality, is observable as functioning in the grand universe. If these findings are dependable, we may conclude that the universes now evolving in outer space are at the present time wholly nonspiritual. And if this is true, it would satisfactorily explain why spirit-endowed beings are in possession of little or no information about these vast

energy manifestations aside from knowing the fact of their physical existence.

3. *Mind Gravity.* By these same principles of comparative computation these experts have attacked the problem of mind-gravity presence and response. The mind unit of estimation was arrived at by averaging three material and three spiritual types of mentality, although the type of mind found in the power directors and their associates proved to be a disturbing factor in the effort to arrive at a basic unit for mind-gravity estimation. There was little to impede the estimation of the present capacity of the Third Source and Center for mind-gravity function in accordance with this theory of totality. Although the findings in this instance are not so conclusive as in the estimates of physical and spirit gravity, they are, comparatively considered, very instructive, even intriguing. These investigators deduce that about eighty-five per cent of the mind-gravity response to the intellectual drawing of the Conjoint Actor takes origin in the existing grand universe. This would suggest the possibility that mind activities are involved in connection with the observable physical activities now in progress throughout the realms of outer space. While this estimate is probably far from accurate, it accords, in principle, with our belief that intelligent force organizers are at present directing universe evolution in the space levels beyond the present outer limits of the grand universe. Whatever the nature of this postulated intelligence, it is apparently not spirit-gravity responsive.

But all these computations are at best estimates based on assumed laws. We think they are fairly reliable. Even if a few spirit beings were located in outer space, their collective presence would not markedly influence calculations involving such enormous measurements.

Personality Gravity is noncomputable. We recognize the circuit, but we cannot measure either qualitative or quantitative realities responsive thereto.

4. Space and Motion

All units of cosmic energy are in primary revolution, are engaged in the execution of their mission, while swinging around the universal orbit. The universes of space and their component systems and worlds are all revolving spheres, moving along the endless circuits of the master universe space levels. Absolutely nothing is stationary in all the master universe except the very center of Havona, the eternal Isle of Paradise, the center of gravity.

The Unqualified Absolute is functionally limited to space, but we are not so sure about the relation of this Absolute to motion. Is motion inherent therein? We do not know. We know that motion is not inherent in space; even the motions *of* space are not innate. But we are not so sure about the relation of the Unqualified to motion. Who, or what, is really responsible for the gigantic activities of force-energy transmutations now in progress out beyond the borders of the present seven superuniverses? Concerning the origin of motion we have the following opinions:

1. We think the Conjoint Actor initiates motion *in* space.

2. If the Conjoint Actor produces the motions *of* space, we cannot prove it.

3. The Universal Absolute does not originate initial motion but does equalize and control all of the tensions originated by motion.

In outer space the force organizers are apparently responsible for the production of the gigantic universe wheels which are now in process of stellar evolution, but their ability so to function must have been made possible by some modification of the space presence of the Unqualified Absolute.

Space is, from the human viewpoint, nothing—negative; it exists only as related to something positive and nonspatial. Space is, however, real. It contains and conditions motion. It even moves. Space motions may be roughly classified as follows:

1. Primary motion—space respiration, the motion of space itself.

2. Secondary motion—the alternate directional swings of the successive space levels.

3. Relative motions—relative in the sense that they are not evaluated with Paradise as a base point. Primary and secondary motions are absolute, motion in relation to unmoving Paradise.

4. Compensatory or correlating movement designed to co-ordinate all other motions.

The present relationship of your sun and its associated planets, while disclosing many relative and absolute motions in space, tends to convey the impression to astronomic observers that you are comparatively stationary in space, and that the surrounding starry clusters and streams are engaged in outward flight at ever-increasing velocities as your calculations proceed outward in space. But such is not the case. You fail to recognize the present outward and uniform expansion of the physical creations of all pervaded space. Your own local creation (Nebadon) participates in this movement of universal outward expansion. The entire seven superuniverses participate in the two-billion-year cycles of space respiration along with the outer regions of the master universe.

When the universes expand and contract, the material masses in pervaded space alternately move against and with the pull of Paradise gravity. The work that is done in moving the material energy mass of creation is *space* work but not *power-energy* work.

Although your spectroscopic estimations of astronomic velocities are fairly reliable when applied to the starry realms belonging to your superuniverse and its associate superuniverses, such reckonings with reference to the realms of outer space are wholly unreliable. Spectral lines are displaced from the normal towards the violet by an approaching star; likewise these lines are displaced towards the red by a receding star. Many influences interpose to make it appear that the recessional velocity of the external universes increases at the rate of more than one hundred miles a second for every million light-years increase in distance. By this method of reckoning, subsequent to the perfection of more powerful telescopes, it will appear that these far-distant systems are in flight from this part of the universe at the unbelievable rate of more than thirty thousand miles a second. But this apparent speed of recession is not real; it results from numerous factors of error embracing angles of observation and other time-space distortions.

But the greatest of all such distortions arises because the vast universes of outer space in the realms next to the domains of the seven superuniverses, seem to be revolving in a direction opposite to that of the grand universe. That is, these myriads of nebulae and their accompanying suns and spheres are at the present time revolving clockwise about the central creation. The seven super-universes revolve about Paradise in a counterclockwise direction. It appears that the second outer universe of galaxies, like the seven superuniverses, revolves counterclockwise about Paradise. And the astronomic observers of Uversa think they detect evidence of revolutionary movements in a third outer belt of far-distant space which are beginning to exhibit directional tendencies of a clockwise nature.

It is probable that these alternate directions of successive space processions of the universes have something to do with the intra-master universe gravity technique of the Universal Absolute, which consists of a co-ordination of forces and an equalization of space tensions. Motion as well as space is a complement or equilibrant of gravity.

5. Space and Time

Like space, time is a bestowal of Paradise, but not in the same sense, only indirectly. Time comes by virtue of motion and because mind is inherently aware of sequentiality. From a practical viewpoint, motion is essential to time, but there is no universal time unit based on motion except in so far as the Paradise-Havona standard day is arbitrarily so recognized. The totality of space respiration destroys its local value as a time source.

Space is not infinite, even though it takes origin from Paradise; not absolute, for it is pervaded by the Unqualified Absolute. We do not know the absolute limits of space, but we do know that the absolute of time is eternity.

Time and space are inseparable only in the time-space creations, the seven superuniverses. Nontemporal space (space without time) theoretically exists, but the only truly nontemporal place is Paradise *area*. Nonspatial time (time without space) exists in mind of the Paradise level of function.

The relatively motionless midspace zones impinging on Paradise and separating pervaded from unpervaded space are the transition zones from time to eternity, hence the necessity of Paradise pilgrims becoming unconscious during this transit when it is to culminate in Paradise citizenship. Time-conscious *visitors* can go to Paradise without thus sleeping, but they remain creatures of time.

Relationships to time do not exist without motion in space, but consciousness of time does. Sequentiality can consciousize time even in the absence of motion. Man's mind is less time-bound than space-bound because of the inherent nature of mind. Even during the days of the earth life in the flesh, though man's mind is rigidly space-bound, the creative human imagination is comparatively time free. But time itself is not genetically a quality of mind.

There are three different levels of time cognizance:

1. Mind-perceived time—consciousness of sequence, motion, and a sense of duration.

2. Spirit-perceived time—insight into motion Godward and the awareness of the motion of ascent to levels of increasing divinity.

3. Personality *creates* a unique time sense out of insight into Reality plus a consciousness of presence and an awareness of duration.

Unspiritual animals know only the past and live in the present. Spirit-indwelt man has powers of prevision (insight); he may visualize the future. Only forward-looking and progressive attitudes are personally real. Static ethics and traditional morality are just slightly superanimal. Nor is stoicism a high order of self-realization. Ethics and morals become truly human when they are dynamic and progressive, alive with universe reality.

The human personality is not merely a concomitant of time-and-space events; the human personality can also act as the cosmic cause of such events.

6. Universal Control

The universe is nonstatic. Stability is not the result of inertia but rather the product of balanced energies, co-operative minds, co-ordinated morontias, spirit overcontrol, and personality unification. Stability is wholly and always proportional to divinity.

In the physical control of the master universe the Universal Father exercises priority and primacy through the Isle of Paradise; God is absolute in the spiritual administration of the cosmos in the person of the Eternal Son. Concerning the domains of mind, the Father and the Son function co-ordinately in the Conjoint Actor.

The Third Source and Center assists in the maintenance of the equilibrium and co-ordination of the combined physical and spiritual energies and organizations by the absoluteness of his grasp of the cosmic mind and by the exercise of his inherent and universal physical- and spiritual-gravity complements. Whenever and wherever there occurs a liaison between the material and the spiritual, such a mind phenomenon is an act of the Infinite Spirit. Mind alone can interassociate the physical forces and energies of the material level with the spiritual powers and beings of the spirit level.

In all your contemplation of universal phenomena, make certain that you take into consideration the interrelation of physical, intellectual, and spiritual energies, and that due allowance is made for the unexpected phenomena attendant upon their unification by personality and for the unpredictable phenomena resulting from the actions and reactions of experiential Deity and the Absolutes.

The universe is highly predictable only in the quantitative or gravity-measurement sense; even the primal physical forces are not responsive to linear gravity, nor are the higher mind meanings and true spirit values of ultimate universe realities. Qualitatively, the universe is not highly predictable as regards new associations of forces, either physical, mindal, or spiritual, although many such combinations of energies or forces become partially predictable when subjected to critical observation. When matter, mind, and spirit are unified by creature personality, we are unable fully to predict the decisions of such a freewill being.

All phases of primordial force, nascent spirit, and other nonpersonal ultimates appear to react in accordance with certain relatively stable but unknown laws and are characterized by a latitude of performance and an elasticity of response which are often disconcerting when encountered in the phenomena of a circumscribed and isolated situation. What is the explanation of this unpredictable freedom of reaction disclosed by these emerging universe actualities? These unknown, unfathomable unpredictables—whether pertaining to the behavior of a primordial unit of force, the reaction of an unidentified level of mind, or the phenomenon of a vast preuniverse in the making in the domains of outer space—probably disclose the activities of the Ultimate and the presence-performances of the Absolutes, which antedate the function of all universe Creators.

We do not really know, but we surmise that such amazing versatility and such profound co-ordination signify the presence and performance of the Absolutes, and that such diversity of response in the face of apparently uniform causation discloses the reaction of the Absolutes, not only to the immediate and situational causation, but also to all other related causations throughout the entire master universe.

Individuals have their guardians of destiny; planets, systems, constellations, universes, and superuniverses each have their respective rulers who labor for the good of their domains. Havona and even the grand universe are watched over by those intrusted with such high responsibilities. But who fosters and cares for the fundamental needs of the master universe as a whole, from Paradise to the fourth and outermost space level? Existentially such overcare is probably attributable to the Paradise Trinity, but from an experiential viewpoint the appearance of the post-Havona universes is dependent on:

1. The Absolutes in potential.

2. The Ultimate in direction.

3. The Supreme in evolutionary co-ordination.

4. The Architects of the Master Universe in administration prior to the appearance of specific rulers.

The Unqualified Absolute pervades all space. We are not altogether clear as to the exact status of the Deity and Universal Absolutes, but we know the latter functions wherever the Deity and Unqualified Absolutes function. The Deity Absolute may be universally present but hardly space present. The Ultimate is, or sometime will be, space present to the outer margins of the fourth space level. We doubt that the Ultimate will ever have a space presence beyond the periphery of the master universe, but within this limit the Ultimate is progressively integrating the creative organization of the potentials of the three Absolutes.

7. THE PART AND THE WHOLE

There is operative throughout all time and space and with regard to all reality of whatever nature an inexorable and impersonal law which is equivalent to the function of a cosmic providence. Mercy characterizes God's attitude of love for the individual; impartiality motivates God's attitude toward the total. The will of God does not necessarily prevail in the part—the heart of any one personality—but his will does actually rule the whole, the universe of universes.

In all his dealings with all his beings it is true that the laws of God are not inherently arbitrary. To you, with your limited vision and finite viewpoint, the acts of God must often appear to be dictatorial and arbitrary. The laws of God are merely the habits of God, his way of repeatedly doing things; and he ever does all things well. You observe that God does the same thing in the same way, repeatedly, simply because that is the best way to do that particular thing in a given circumstance; and the best way is the right way, and therefore does infinite wisdom always order it done in that precise and perfect manner. You should also remember that nature is not the exclusive act of Deity; other influences are present in those phenomena which man calls nature.

It is repugnant to the divine nature to suffer any sort of deterioration or ever to permit the execution of any purely personal act in an inferior way. It should be made clear, however, that, *if*, in the divinity of any situation, in the extremity of any circumstance, in any case where the course of supreme wisdom might indicate the demand for different conduct—if the demands of perfection might for any reason dictate another method of reaction, a better one, then and there would the all-wise God function in that better and more suitable way. That would be the expression of a higher law, not the reversal of a lower law.

God is not a habit-bound slave to the chronicity of the repetition of his own voluntary acts. There is no conflict among the laws of the Infinite; they are all perfections of the infallible nature; they are all the unquestioned acts expressive of faultless decisions. Law is the unchanging reaction of an infinite, perfect, and divine mind. The acts of God are all volitional notwithstanding this apparent sameness. In God there "is no variableness neither shadow of changing." But all this which can be truly said of the Universal Father cannot be said with equal certainty of all his subordinate intelligences or of his evolutionary creatures.

Because God is changeless, therefore can you depend, in all ordinary circumstances, on his doing the same thing in the same identical and ordinary way. God is the assurance of stability for all created things and beings. He is God; therefore he changes not.

And all this steadfastness of conduct and uniformity of action is personal, conscious, and highly volitional, for the great God is not a helpless slave to his own perfection and infinity. God is not a self-acting automatic force; he is not a slavish law-bound power. God is neither a mathematical equation nor a chemical formula. He is a freewill and primal personality. He is the Universal Father, a being surcharged with personality and the universal fount of all creature personality.

The will of God does not uniformly prevail in the heart of the God-seeking material mortal, but if the time frame is enlarged beyond the moment to embrace the whole of the first life, then does God's will become increasingly discernible in the spirit fruits which are borne in the lives of the spirit-led children of God. And then, if human life is further enlarged to include the morontia experience, the divine will is observed to shine brighter and brighter in the spiritualizing acts of those creatures of time who have begun to taste the divine delights of experiencing the relationship of the personality of man with the personality of the Universal Father.

The Fatherhood of God and the brotherhood of man present the paradox of the part and the whole on the level of personality. God loves *each* individual as an individual child in the heavenly family. Yet God thus loves *every* individual; he is no respecter of persons, and the universality of his love brings into being a relationship of the whole, the universal brotherhood.

The love of the Father absolutely individualizes each personality as a unique child of the Universal Father, a child without duplicate in infinity, a will creature irreplaceable in all eternity. The Father's love glorifies each child of God, illuminating each member of the celestial family, sharply silhouetting the unique nature of each personal being against the impersonal levels that lie outside the fraternal circuit of the Father of all. The love of God strikingly portrays the transcendent value of each will creature, unmistakably reveals the high value which the Universal Father has placed upon each and every one of his children from the highest creator personality of Paradise status to the lowest personality of will dignity among the savage tribes of men in the dawn of the human species on some evolutionary world of time and space.

This very love of God for the individual brings into being the divine family of all individuals, the universal brotherhood of the freewill children of the Paradise Father. And this brotherhood, being universal, is a relationship of the whole. Brotherhood, when universal, discloses not the *each* relationship, but the *all* relationship. Brotherhood is a reality of the total and therefore discloses qualities of the whole in contradistinction to qualities of the part.

Brotherhood constitutes a fact of relationship between every personality in universal existence. No person can escape the benefits or the penalties that may come as a result of relationship to other persons. The part profits or suffers in measure with the whole. The good effort of each man benefits all men; the error or evil of each man augments the tribulation of all men. As moves the part, so moves the whole. As the progress of the whole, so the progress of the part. The relative velocities of part and whole determine whether the part is retarded by the inertia of the whole or is carried forward by the momentum of the cosmic brotherhood.

It is a mystery that God is a highly personal self-conscious being with residential headquarters, and at the same time personally present in such a vast universe and personally in contact with such a well-nigh infinite number of beings. That such a phenomenon is a mystery beyond human comprehension should not in the least lessen your faith. Do not allow the magnitude of the infinity, the immensity of the eternity, and the grandeur and glory of the matchless character of God to overawe, stagger, or discourage you; for the Father is not very far from any one of you; he dwells within you, and in him do we all literally move, actually live, and veritably have our being.

Even though the Paradise Father functions through his divine creators and his creature children, he also enjoys the most intimate inner contact with you, so sublime, so highly personal, that it is even beyond my comprehension—that mysterious communion of the Father fragment with the human soul and with the mortal mind of its actual indwelling. Knowing what you do of these gifts of God, you therefore know that the Father is in intimate touch, not only with his divine associates, but also with his evolutionary mortal children of time. The Father indeed abides on Paradise, but his divine presence also dwells in the minds of men.

Even though the spirit of a Son be poured out upon all flesh, even though a Son once dwelt with you in the likeness of mortal flesh, even though the seraphim personally guard and guide you, how can any of these divine beings of the Second and Third Centers ever hope to come as near to you or to understand you as fully as the Father, who has given a part of himself to be in you, to be your real and divine, even your eternal, self?

8. Matter, Mind, and Spirit

"God is spirit," but Paradise is not. The material universe is always the arena wherein take place all spiritual activities; spirit beings and spirit ascenders live and work on physical spheres of material reality.

The bestowal of cosmic force, the domain of cosmic gravity, is the function of the Isle of Paradise. All original force-energy proceeds from Paradise, and the matter for the making of untold universes now circulates throughout the master universe in the form of a supergravity presence which constitutes the force-charge of pervaded space.

Whatever the transformations of force in the outlying universes, having gone out from Paradise, it journeys on subject to the never-ending, ever-present, unfailing pull of the eternal Isle, obediently and inherently swinging on forever around the eternal space paths of the universes. Physical energy is the one reality which is true and steadfast in its obedience to universal law. Only in the realms of creature volition has there been deviation from the divine paths and the original plans. Power and energy are the universal evidences of the stability, constancy, and eternity of the central Isle of Paradise.

The bestowal of spirit and the spiritualization of personalities, the domain of spiritual gravity, is the realm of the Eternal Son. And this spirit gravity of the Son, ever drawing all spiritual realities to himself, is just as real and absolute as is the all-powerful material grasp of the Isle of Paradise. But material-minded man is naturally more familiar with the material manifestations of a physical nature than with the equally real and mighty operations of a spiritual nature which are discerned only by the spiritual insight of the soul.

As the mind of any personality in the universe becomes more spiritual—Godlike—it becomes less responsive to material gravity. Reality, measured by physical-gravity response, is the antithesis of reality as determined by quality of spirit content. Physical-gravity action is a quantitative determiner of nonspirit energy; spiritual-gravity action is the qualitative measure of the living energy of divinity.

What Paradise is to the physical creation, and what the Eternal Son is to the spiritual universe, the Conjoint Actor is to the realms of mind—the intelligent universe of material, morontial, and spiritual beings and personalities.

The Conjoint Actor reacts to both material and spiritual realities and therefore inherently becomes the universal minister to all intelligent beings, beings who may represent a union of both the material and spiritual phases of creation. The endowment of intelligence, the ministry to the material and the spiritual in the phenomenon of mind, is the exclusive domain of the Conjoint Actor, who thus becomes the partner of the spiritual mind, the essence of the morontia mind, and the substance of the material mind of the evolutionary creatures of time.

Mind is the technique whereby spirit realities become experiential to creature personalities. And in the last analysis the unifying possibilities of even human mind, the ability to co-ordinate things, ideas, and values, is supermaterial.

Though it is hardly possible for the mortal mind to comprehend the seven levels of relative cosmic reality, the human intellect should be able to grasp much of the meaning of three functioning levels of finite reality:

1. *Matter.* Organized energy which is subject to linear gravity except as it is modified by motion and conditioned by mind.

2. *Mind.* Organized consciousness which is not wholly subject to material gravity, and which becomes truly liberated when modified by spirit.

3. *Spirit.* The highest personal reality. True spirit is not subject to physical gravity but eventually becomes the motivating influence of all evolving energy systems of personality dignity.

The goal of existence of all personalities is spirit; material manifestations are relative, and the cosmic mind intervenes between these universal opposites. The bestowal of mind and the ministration of spirit are the work of the associate persons of Deity, the Infinite Spirit and the Eternal Son. Total Deity reality is not mind but spirit-mind—mind-spirit unified by personality. Nevertheless the absolutes of both the spirit and the thing converge in the person of the Universal Father.

On Paradise the three energies, physical, mindal, and spiritual, are coordinate. In the evolutionary cosmos energy-matter is dominant except in personality, where spirit, through the mediation of mind, is striving for the mastery. Spirit is the fundamental reality of the personality experience of all creatures because God is spirit. Spirit is unchanging, and therefore, in all personality relations, it transcends both mind and matter, which are experiential variables of progressive attainment.

In cosmic evolution matter becomes a philosophic shadow cast by mind in the presence of spirit luminosity of divine enlightenment, but this does not invalidate the reality of matter-energy. Mind, matter, and spirit are equally real, but they are not of equal value to personality in the attainment of divinity. Consciousness of divinity is a progressive spiritual experience.

The brighter the shining of the spiritualized personality (the Father in the universe, the fragment of potential spirit personality in the individual creature), the greater the shadow cast by the intervening mind upon its material investment. In time, man's body is just as real as mind or spirit, but in death, both mind (identity) and spirit survive while the body does not. A cosmic reality can be nonexistent in personality experience. And so your Greek figure of speech—the material as the shadow of the more real spirit substance—does have a philosophic significance.

9. Personal Realities

Spirit is the basic personal reality in the universes, and personality is basic to all progressing experience with spiritual reality. Every phase of personality experience on every successive level of universe progression swarms with clues to the discovery of alluring personal realities. Man's true destiny consists in the creation of new and spirit goals and then in responding to the cosmic allurements of such supernal goals of nonmaterial value.

Love is the secret of beneficial association between personalities. You cannot really know a person as the result of a single contact. You cannot appreciatingly know music through mathematical deduction, even though music is a form of mathematical rhythm. The number assigned to a telephone subscriber does not in any manner identify the personality of that subscriber or signify anything concerning his character.

Mathematics, material science, is indispensable to the intelligent discussion of the material aspects of the universe, but such knowledge is not necessarily a part of the higher realization of truth or of the personal appreciation of spiritual realities. Not only in the realms of life but even in the world of physical energy, the sum of two or more things is very often something *more* than, or something *different* from, the predictable additive consequences of such unions. The entire science of mathematics, the whole domain of philosophy, the highest physics or chemistry, could not predict or know that the union of two gaseous hydrogen atoms with one gaseous oxygen atom would result in a new and qualitatively superadditive substance—liquid water. The understanding knowledge of this one physiochemical phenomenon should have prevented the development of materialistic philosophy and mechanistic cosmology.

Technical analysis does not reveal what a person or a thing can do. For example: Water is used effectively to extinguish fire. That water will put out fire is a fact of everyday experience, but no analysis of water could ever be made to disclose such a property. Analysis determines that water is composed of hydrogen and oxygen; a further study of these elements discloses that oxygen is the real supporter of combustion and that hydrogen will itself freely burn.

Your religion is becoming real because it is emerging from the slavery of fear and the bondage of superstition. Your philosophy struggles for emancipation from dogma and tradition. Your science is engaged in the agelong contest between truth and error while it fights for deliverance from the bondage of abstraction, the slavery of mathematics, and the relative blindness of mechanistic materialism.

Mortal man has a spirit nucleus. The mind is a personal-energy system existing around a divine spirit nucleus and functioning in a material environment. Such a living relationship of personal mind and spirit constitutes the universe potential of eternal personality. Real trouble, lasting disappointment, serious defeat, or inescapable death can come only after self-concepts presume fully to displace the governing power of the central spirit nucleus, thereby disrupting the cosmic scheme of personality identity.

[Presented by a Perfector of Wisdom acting by authority of the Ancients of Days.]

THREE

THE SACRED SPHERES
OF PARADISE

(PAPER 13)

BETWEEN the central Isle of Paradise and the innermost of
the Havona planetary circuits there are situated in space three
lesser circuits of special spheres. The innermost circuit consists of
the seven secret spheres of the Universal Father; the second group
is composed of the seven luminous worlds of the Eternal Son; in the
outermost are the seven immense spheres of the Infinite Spirit, the
executive-headquarters worlds of the Seven Master Spirits.

These three seven-world circuits of the Father, the Son, and
the Spirit are spheres of unexcelled grandeur and unimagined glory.
Even their material or physical construction is of an order unrevealed to you. Each circuit is diverse in material, and each world
of each circuit is different excepting the seven worlds of the Son,
which are alike in physical constitution. All twenty-one are enormous spheres, and each group of seven is differently eternalized. As
far as we know they have always been; like Paradise they are eternal.
There exists neither record nor tradition of their origin.

The seven secret spheres of the Universal Father, circulating about Paradise in close proximity to the eternal Isle, are highly
reflective of the spiritual luminosity of the central shining of the
eternal Deities, shedding this light of divine glory throughout Paradise and even upon the seven circuits of Havona.

On the seven sacred worlds of the Eternal Son there appear to take origin the impersonal energies of spirit luminosity. No personal being may sojourn on any of these seven shining realms. With spiritual glory they illuminate all Paradise and Havona, and they directionize pure spirit luminosity to the seven superuniverses. These brilliant spheres of the second circuit likewise emit their light (light without heat) to Paradise and to the billion worlds of the seven-circuited central universe.

The seven worlds of the Infinite Spirit are occupied by the Seven Master Spirits, who preside over the destinies of the seven superuniverses, sending forth the spiritual illumination of the Third Person of Deity to these creations of time and space. And all Havona, but not the Isle of Paradise, is bathed in these spiritualizing influences.

Although the worlds of the Father are ultimate status spheres for all Father-endowed personalities, this is not their exclusive function. Many beings and entities other than personal sojourn on these worlds. Each world in the circuit of the Father and the circuit of the Spirit has a distinct type of permanent citizenship, but we think the Son's worlds are inhabited by uniform types of other-than-personal beings. Father fragments are among the natives of Divinington; the other orders of permanent citizenship are unrevealed to you.

The twenty-one Paradise satellites serve many purposes in both central and superuniverses not disclosed in these narratives. You are able to understand so little of the life of these spheres that you cannot hope to gain anything like a consistent view of them, either as to nature or function; thousands of activities are there going on which are unrevealed to you. These twenty-one spheres embrace the *potentials* of the function of the master universe. These papers afford only a fleeting glimpse of certain circumscribed activities pertaining to the pres-ent universe age of the grand universe—rather, one of the seven sectors of the grand universe.

1. THE SEVEN SACRED WORLDS OF THE FATHER

The Father's circuit of sacred life spheres contains the only inherent personality secrets in the universe of universes. These satellites of Paradise, the innermost of the three circuits, are the only forbidden domains concerned with personality in the central universe. Nether Paradise and the worlds of the Son are likewise closed to personalities, but neither of those realms is in any way directly concerned with personality.

The Paradise worlds of the Father are directed by the highest order of the Stationary Sons of the Trinity, the Trinitized Secrets of Supremacy. Of these worlds I can tell little; of their manifold activities I may tell less. Such information concerns only those beings who function thereon and go forth therefrom. And though I am somewhat familiar with six of these special worlds, never have I landed on Divinington; that world is wholly forbidden to me.

One of the reasons for the secrecy of these worlds is because each of these sacred spheres enjoys a specialized representation, or manifestation, of the Deities composing the Paradise Trinity; not a personality, but a unique presence of Divinity which can only be appreciated and comprehended by those particular groups of intelligences resident on, or admissible to, that particular sphere. The Trinitized Secrets of Supremacy are the personal agents of these specialized and impersonal presences of Divinity. And the Secrets of Supremacy are highly personal beings, superbly endowed and marvelously adapted to their exalted and exacting work.

1. DIVININGTON. This world is, in a unique sense, the "bosom of the Father," the personal-communion sphere of the Universal Father, and thereon is a special manifestation of his divinity. Divinington is the Paradise rendezvous of the Thought Adjusters, but it is also the home of numerous other entities, personalities, and other beings taking origin in the Universal Father. Many personalities besides the Eternal Son are of direct origin by the solitary acts of the Universal Father. Only the Father fragments and those personalities and other beings of direct and exclusive origin in the Universal Father fraternize and function on this abode.

The secrets of Divinington include the secret of the bestowal and mission of Thought Adjusters. Their nature, origin, and the technique of their contact with the lowly creatures of the evolutionary worlds is a secret of this Paradise sphere. These amazing transactions do not personally concern the rest of us, and therefore do the Deities deem it proper to withhold certain features of this great and divine ministry from our full understanding. In so far as we come in contact with this phase of divine activity, we are permitted full knowledge of these transactions, but concerning the intimate details of this great bestowal we are not fully informed.

This sphere also holds the secrets of the nature, purpose, and activities of all other forms of Father fragments, of the Gravity Messengers, and of hosts of other beings unrevealed to you. It is highly probable that those truths pertaining to Divinington which are withheld from me, if revealed, would merely confuse and handicap me in my present work, and still again, perhaps they are beyond the conceptual capacity of my order of being.

2. SONARINGTON. This sphere is the "bosom of the Son," the personal receiving world of the Eternal Son. It is the Paradise headquarters of the descending and ascending Sons of God when, and after, they are fully accredited and finally approved. This world is the Paradise home for all Sons of the Eternal Son and of his co-ordinate and associate Sons. There are numerous orders of divine sonship attached to this supernal abode which have not been revealed to mortals since they are not concerned with the plans of the ascension scheme of human spiritual progression through the universes and on to Paradise.

The secrets of Sonarington include the secret of the incarnation of the divine Sons. When a Son of God becomes a Son of Man, is literally born of woman, as occurred on your world nineteen hundred years ago, it is a universal mystery. It is occurring right along throughout the universes, and it is a Sonarington secret of divine sonship. The Adjusters are a mystery of God the Father. The incarnation of the divine Sons is a mystery of God the Son; it is a secret locked up in the seventh sector of Sonarington, a realm penetrated by none save those who have personally passed through this unique experience. Only those phases of incarnation having to do with your ascension career have been brought to your notice. There are many other phases of the mystery of the incarnation of the Paradise Sons of unrevealed types on missions of universe service which are undisclosed to you. And there are still other Sonarington mysteries.

3. SPIRITINGTON. This world is the "bosom of the Spirit," the Paradise home of the high beings that exclusively represent the Infinite Spirit. Here forgather the Seven Master Spirits and certain of their offspring from all universes. At this celestial abode may also be

found numerous unrevealed orders of spirit personalities, beings assigned to the manifold activities of the universe not associated with the plans of upstepping the mortal creatures of time to the Paradise levels of eternity.

The secrets of Spiritington involve the impenetrable mysteries of reflectivity. We tell you of the vast and universal phenomenon of reflectivity, more particularly as it is operative on the headquarters worlds of the seven superuniverses, but we never fully explain this phenomenon, for we do not fully understand it. Much, very much, we do comprehend, but many basic details are still mysterious to us. Reflectivity is a secret of God the Spirit. You have been instructed concerning reflectivity functions in relation to the ascension scheme of mortal survival, and it does so operate, but reflectivity is also an indispensable feature of the normal working of numerous other phases of universe occupation. This endowment of the Infinite Spirit is also utilized in channels other than those of intelligence gathering and information dissemination. And there are other secrets of Spiritington.

4. VICEGERINGTON. This planet is the "bosom of the Father and the Son" and is the secret sphere of certain unrevealed beings who take origin by the acts of the Father and the Son. This is also the Paradise home of many glorified beings of complex ancestry, those whose origin is complicated because of the many diverse techniques operative in the seven superuniverses. Many groups of beings forgather on this world whose identity has not been revealed to Urantia mortals.

The secrets of Vicegerington include the secrets of trinitization, and trinitization constitutes the secret of authority to represent the Trinity, to act as vicegerents of the Gods. Authority to represent the Trinity attaches only to those beings, revealed and unrevealed, who are trinitized, created, eventuated, or eternalized by any two or all three of the Paradise Trinity. Personalities brought into being by the trinitizing acts of certain types of glorified creatures represent no more than the conceptual potential mobilized in that trinitization, albeit such creatures may ascend the path of Deity embrace open to all of their kind.

Nontrinitized beings do not fully understand the technique of trinitization by either two or three Creators or by certain creatures. You will never fully understand such a phenomenon unless, in the far-distant future of your glorified career, you should essay and succeed in such an adventure, because otherwise these secrets of Vicegerington will always be forbidden you. But to me, a high Trinity-origin being, all sectors of Vicegerington are open. I fully understand, and just as fully and sacredly protect, the secret of my origin and destiny.

There are still other forms and phases of trinitization which have not been brought to the notice of the Urantia peoples, and these experiences, in their personal aspects, are duly protected in the secret sector of Vicegerington.

5. SOLITARINGTON. This world is the "bosom of the Father and the Spirit" and is the rendezvous of a magnificent host of unrevealed beings of origin in the conjoint acts of the Universal Father and the Infinite Spirit, beings who partake of the traits of the Father in addition to their Spirit inheritance.

This is also the home of the Solitary Messengers and of other personalities of the superangelic orders. You know of very few of these beings; there are vast numbers of orders unrevealed on Urantia. Because they are domiciled on the fifth world, it does not necessarily follow that the Father had aught to do with the creation of Solitary Messengers or their superangelic associates, but in this universe age he does have to do with their function. During the present universe age this is also the status sphere of the Universe Power Directors.

There are numerous additional orders of spirit personalities, beings unknown to mortal man, who look upon Solitarington as their Paradise home sphere. It should be remembered that all divisions and levels of universe activities are just as fully provided with spirit ministers as is the realm concerned with helping mortal man ascend to his divine Paradise destiny.

The secrets of Solitarington. Besides certain secrets of trinitization, this world holds the secrets of the personal relation of the Infinite Spirit with certain of the higher offspring of the Third Source and Center. On Solitarington are held the mysteries of the intimate association of numerous unrevealed orders with the spirits of the Father, of the Son, and of the Spirit, with the threefold spirit of the Trinity, and with the spirits of the Supreme, the Ultimate, and the Supreme-Ultimate.

6. SERAPHINGTON. This sphere is the "bosom of the Son and the Spirit" and is the home world of the vast hosts of unrevealed beings created by the Son and the Spirit. This is also the destiny sphere of all ministering orders of the angelic hosts, including supernaphim, seconaphim, and seraphim. There also serve in the central and outlying universes many orders of superb spirits who are not "ministering spirits to those who shall be heirs of salvation." All these spirit workers in all levels and realms of universe activities look upon Seraphington as their Paradise home.

The secrets of Seraphington involve a threefold mystery, only one of which I may mention—the mystery of seraphic transport. The ability of various orders of seraphim and allied spirit beings to envelop within their spirit forms all orders of nonmaterial personalities and to carry them away on lengthy interplanetary journeys, is a secret locked up in the sacred sectors of Seraphington. The transport seraphim comprehend this mystery, but they do not communicate it to the rest of us, or perhaps they cannot. The other mysteries of Seraphington pertain to the personal experiences of types of spirit servers as yet not revealed to mortals. And we refrain from discussing the secrets of such closely related beings because you can almost comprehend such near orders of existence, and it would be akin to betrayal of trust to present even our partial knowledge of such phenomena.

7. ASCENDINGTON. This unique world is the "bosom of the Father, Son, and Spirit," the rendezvous of the ascendant creatures of space, the receiving sphere of the pilgrims of time who are passing through the Havona universe on their way to Paradise. Ascendington is the actual Paradise home of the ascendant souls of time and space until they attain Paradise status. You mortals will spend most of your Havona "vacations" on Ascendington. During your Havona life Ascendington will be to you what the reversion directors were during the local and superuniverse ascension. Here you will engage in thousands of activities which are beyond the grasp of mortal imagination. And as on every previous advance in the Godward ascent, your human self will here enter into new relationships with your divine self.

The secrets of Ascendington include the mystery of the gradual and certain building up in the material and mortal mind of a spiritual and potentially immortal counterpart of character and identity. This phenomenon constitutes one of the most perplexing mysteries of the universes—the evolution of an immortal soul within the mind of a mortal and material creature.

You will never fully understand this mysterious transaction until you reach Ascendington. And that is just why all Ascendington will be open to your wondering gaze. One seventh of Ascendington is forbidden to me—that sector concerned with this very secret which is (or will be) the exclusive experience and possession of your type of being. This experience belongs to your human order of existence. My order of personality is not directly concerned with such transactions. It is therefore forbidden to me and eventually revealed to you. But even after it is revealed to you, for some reason it forever remains your secret. You do not reveal it to us nor to any other order of beings. We know about the eternal fusion of a divine Adjuster and an immortal soul of human origin, but the ascendant finaliters know this very experience as an absolute reality.

2. FATHER-WORLD RELATIONSHIPS

These home worlds of the diverse orders of spiritual beings are tremendous and stupendous spheres, and they are equal to Paradise in their matchless beauty and superb glory. They are rendezvous worlds, reunion spheres, serving as permanent cosmic addresses. As finaliters you will be domiciled on Paradise, but Ascendington will be your home address at all times, even when you enter service in outer space. Through all eternity you will regard Ascendington as your home of sentimental memories and reminiscent recollections. When you become seventh-stage spirit beings, possibly you will give up your residential status on Paradise.

If outer universes are in the making, if they are to be inhabited by time creatures of ascension potential, then we infer that these children of the future will also be destined to look upon Ascendington as their Paradise home world.

Ascendington is the only sacred sphere that will be unreservedly open to your inspection as a Paradise arrival. Vicegerington is the only sacred sphere that is wholly and unreservedly open to my scrutiny. Though its secrets are concerned in my origin, in this universe age I do not regard Vicegerington as my home. Trinity-origin beings and trinitized beings are not the same.

The Trinity-origin beings do not fully share the Father's worlds; they have their sole homes on the Isle of Paradise in close proximity to the Most Holy Sphere. They often appear on Ascendington, the "bosom of the Father-Son-Spirit," where they fraternize with their brethren who have come up from the lowly worlds of space.

You might assume that Creator Sons, being of Father-Son origin, would regard Vicegerington as their home, but such is not the case in this universe age of the function of God the Sevenfold. And there are many similar problems that will perplex you, for you are sure to encounter many difficulties as you attempt to understand these things which are so near Paradise. Nor can you successfully reason out these questions; you know so little. And if you knew more about the Father's worlds, you would simply encounter more difficulties until you knew *all* about them. Status on any of these secret worlds is acquired by service as well as by nature of origin, and the successive universe ages may and do redistribute certain of these personality groupings.

The worlds of the inner circuit are really fraternal or status worlds more than actual residential spheres. Mortals will attain some status on each of the Father's worlds save one. For example: When you mortals attain Havona, you are granted clearance for Ascendington, where you are most welcome, but you are not permitted to visit the other six sacred worlds. Subsequent to your passage through the Paradise regime and after your admission to the Corps of the Finality, you are granted clearance for Sonarington since you are sons of God as well as ascenders—and you are even more. But there will always remain one seventh of Sonarington, the sector of the incarnation secrets of the divine Sons, which will not be open to your scrutiny. Never will those secrets be revealed to the ascendant sons of God.

Eventually you will have full access to Ascendington and relative access to the other spheres of the Father except Divinington. But even when you are granted permission to land on five additional secret spheres, after you have become a finaliter, you will not be allowed to visit all sectors of such worlds. Nor will you be permitted to land on the shores of Divinington, the "bosom of the Father," though you shall surely stand repeatedly at the "right hand of the Father." Never throughout all eternity will there arise any necessity for your presence on the world of the Thought Adjusters.

These rendezvous worlds of spirit life are forbidden ground to the extent that we are asked not to negotiate entrance to those phases of these spheres which are wholly outside our realms of experience. You may become creature perfect even as the Universal Father is deity perfect, but you may not know all the experiential secrets of all other orders of universe personalities. When the Creator has an experiential personality secret with his creature, the Creator preserves that secret in eternal confidence.

All these secrets are supposedly known to the collective body of the Trinitized Secrets of Supremacy. These beings are fully known only by their special world groups; they are little comprehended by other orders. After you attain Paradise, you will know and ardently love the ten Secrets of Supremacy who direct Ascendington. Excepting Divinington, you will also achieve a partial understanding of the Secrets of Supremacy on the other worlds of the Father, though not so perfectly as on Ascendington.

The Trinitized Secrets of Supremacy, as their name might suggest, are related to the Supreme; they are likewise related to the Ultimate and to the future Supreme-Ultimate. These Secrets of Supremacy are the secrets of the Supreme and also the secrets of the Ultimate, even the secrets of the Supreme-Ultimate.

3. The Sacred Worlds of the Eternal Son

The seven luminous spheres of the Eternal Son are the worlds of the seven phases of pure-spirit existence. These shining orbs are the source of the threefold light of Paradise and Havona, their influence being largely, but not wholly, confined to the central universe.

Personality is not present on these Paradise satellites; therefore is there little concerning these pure-spirit abodes which can be presented to the mortal and material personality. We are taught that these worlds teem with the otherwise-than-personal life of the beings of the Eternal Son. We infer that these entities are being assembled for ministry in the projected new universes of outer space. The Paradise philosophers maintain that each Paradise cycle, about two billion years of Urantia time, witnesses the creation of additional reserves of these orders on the secret worlds of the Eternal Son.

As far as I am informed, no personality has ever been on any one of these spheres of the Eternal Son. I have never been assigned to visit one of these worlds in all my long experience in and out of Paradise. Even the personalities cocreated by the Eternal Son do not go to these worlds. We infer that all types of impersonal spirits—regardless of parentage—are admitted to these spirit homes. As I am a person and have a spirit form, no doubt such a world would seem empty and deserted even if I were permitted to pay it a visit. High spirit personalities are not given to the gratification of purposeless curiosity, purely useless adventure. There is at all times altogether too much intriguing and purposeful adventure to permit the development of any great interest in those projects which are either futile or unreal.

4. THE WORLDS OF THE INFINITE SPIRIT

Between the inner circuit of Havona and the shining spheres of the Eternal Son there circle the seven orbs of the Infinite Spirit, worlds inhabited by the offspring of the Infinite Spirit, by the trinitized sons of glorified created personalities, and by other types of unrevealed beings concerned with the effective administration of the many enterprises of the various realms of universe activities.

The Seven Master Spirits are the supreme and ultimate representatives of the Infinite Spirit. They maintain their personal stations, their power focuses, on the periphery of Paradise, but all operations concerned with their management and direction of the grand universe are conducted on and from these seven special executive spheres of the Infinite Spirit. The Seven Master Spirits are, in reality, the mind-spirit balance wheel of the universe of universes, an all-embracing, all-encompassing, and all-co-ordinating power of central location.

From these seven special spheres the Master Spirits operate to equalize and stabilize the cosmic-mind circuits of the grand universe. They also have to do with the differential spiritual attitude and presence of the Deities throughout the grand universe. Physical reactions are uniform, unvarying, and always instantaneous and automatic. But experiential spiritual presence is in accordance with the underlying conditions or states of spiritual receptivity inherent in the individual minds of the realms.

Physical authority, presence, and function are unvarying in all the universes, small or great. The differing factor in spiritual presence, or reaction, is the fluctuating differential in its recognition and reception by will creatures. Whereas the spiritual presence of absolute and existential Deity is in no manner whatever influenced by attitudes of loyalty or disloyalty on the part of created beings, at the same time it is true that the functioning presence of subabsolute and experiential Deity is definitely and directly influenced by the decisions, choices, and will-attitudes of such finite creature beings—by the loyalty and devotion of the individual being, planet, system, constellation, or universe. But this spiritual presence of divinity is not whimsical nor arbitrary; its experiential variance is inherent in the freewill endowment of personal creatures.

The determiner of the differential of spiritual presence exists in your own hearts and minds and consists in the manner of your own choosing, in the decisions of your minds, and in the determination of your own wills. This differential is inherent in the freewill reactions of intelligent personal beings, beings whom the Universal Father has ordained shall exercise this liberty of choosing. And the Deities are ever true to the ebb and flow of their spirits in meeting and satisfying the conditions and demands of this differential of creature choice, now bestowing more of their presence in response to a sincere desire for the same and again withdrawing themselves from the scene as their creatures decide adversely in the exercise of their divinely bestowed freedom of choice. And thus does the spirit of divinity become humbly obedient to the choosing of the creatures of the realms.

The executive abodes of the Seven Master Spirits are, in reality, the Paradise headquarters of the seven superuniverses and their correlated segments in outer space. Each Master Spirit presides over one superuniverse, and each of these seven worlds is exclusively assigned to one of the Master Spirits. There is literally no phase of the sub-Paradise administration of the seven superuniverses which is not provided for on these executive worlds. They are not so exclusive as the spheres of the Father or those of the Son, and though residential status is limited to native beings and those who work thereon, these seven administrative planets are always open to all beings who desire to visit them, and who can command the necessary means of transit.

To me, these executive worlds are the most interesting and intriguing spots outside of Paradise. In no other place in the wide universe can one observe such varied activities, involving so many different orders of living beings, having to do with operations on so many diverse levels, occupations at once material, intellectual, and spiritual. When I am accorded a period of release from assignment, if I chance to be on Paradise or in Havona, I usually proceed to one of these busy worlds of the Seven Master Spirits, there to inspire my mind with such spectacles of enterprise, devotion, loyalty, wisdom, and effectiveness. Nowhere else can I observe such an amazing interassociation of personality performances on all seven levels of universe reality. And I am always stimulated by the activities of those who well know how to do their work, and who so thoroughly enjoy doing it.

[Presented by a Perfector of Wisdom commissioned thus to function by the Ancients of Days on Uversa.]

FOUR

THE CENTRAL
AND DIVINE UNIVERSE

(PAPER 14)

THE perfect and divine universe occupies the center of all creation; it is the eternal core around which the vast creations
of time and space revolve. Paradise is the gigantic nuclear Isle of
absolute stability which rests motionless at the very heart of the
magnificent eternal universe. This central planetary family is called
Havona and is far-distant from the local universe of Nebadon. It is of
enormous dimensions and almost unbelievable mass and consists of
one billion spheres of unimagined beauty and superb grandeur, but
the true magnitude of this vast creation is really beyond the understanding grasp of the human mind.

   This is the one and only settled, perfect, and established aggregation of worlds. This is a wholly created and perfect universe; it is
not an evolutionary development. This is the eternal core of perfection, about which swirls that endless procession of universes which
constitute the tremendous evolutionary experiment, the audacious
adventure of the Creator Sons of God, who aspire to duplicate in
time and to reproduce in space the pattern universe, the ideal of
divine completeness, supreme finality, ultimate reality, and eternal
perfection.

1. THE PARADISE-HAVONA SYSTEM

From the periphery of Paradise to the inner borders of the seven superuniverses there are the following seven space conditions and motions:

1. The quiescent midspace zones impinging on Paradise.

2. The clockwise processional of the three Paradise and the seven Havona circuits.

3. The semiquiet space zone separating the Havona circuits from the dark gravity bodies of the central universe.

4. The inner, counterclockwise-moving belt of the dark gravity bodies.

5. The second unique space zone dividing the two space paths of the dark gravity bodies.

6. The outer belt of dark gravity bodies, revolving clockwise around Paradise.

7. A third space zone—a semiquiet zone—separating the outer belt of dark gravity bodies from the innermost circuits of the seven superuniverses.

The billion worlds of Havona are arranged in seven concentric circuits immediately surrounding the three circuits of Paradise satellites. There are upwards of thirty-five million worlds in the innermost Havona circuit and over two hundred and forty-five million in the outermost, with proportionate numbers intervening. Each circuit differs, but all are perfectly balanced and exquisitely organized, and each is pervaded by a specialized representation of the Infinite Spirit, one of the Seven Spirits of the Circuits. In addition to other functions this impersonal Spirit co-ordinates the conduct of celestial affairs throughout each circuit.

The Havona planetary circuits are not superimposed; their worlds follow each other in an orderly linear procession. The central universe whirls around the stationary Isle of Paradise in one vast plane, consisting of ten concentric stabilized units—the three circuits of Paradise spheres and the seven circuits of Havona worlds. Physically regarded, the Havona and the Paradise circuits are all one and the same system; their separation is in recognition of functional and administrative segregation.

Time is not reckoned on Paradise; the sequence of successive events is inherent in the concept of those who are indigenous to the central Isle. But time is germane to the Havona circuits and to numerous beings of both celestial and terrestrial origin sojourning thereon. Each Havona world has its own local time, determined by its circuit. All worlds in a given circuit have the same length of year since they uniformly swing around Paradise, and the length of these planetary years decreases from the outermost to the innermost circuit.

Besides Havona-circuit time, there is the Paradise-Havona standard day and other time designations which are determined on, and are sent out from, the seven Paradise satellites of the Infinite Spirit. The Paradise-Havona standard day is based on the length of time required for the planetary abodes of the first or inner Havona circuit to complete one revolution around the Isle of Paradise; and though their velocity is enormous, owing to their situation between the dark gravity bodies and gigantic Paradise, it requires almost one thousand years for these spheres to complete their circuit. You have unwittingly read the truth when your eyes rested on the statement "A day is as a thousand years with God, as but a watch in the night." One Paradise-Havona day is just seven minutes, three and one-eighth seconds less than one thousand years of the present Urantia leap-year calendar.

This Paradise-Havona day is the standard time measurement for the seven superuniverses, although each maintains its own internal time standards.

On the outskirts of this vast central universe, far out beyond the seventh belt of Havona worlds, there swirl an unbelievable number of enormous dark gravity bodies. These multitudinous dark masses are quite unlike other space bodies in many particulars; even in form they are very different. These dark gravity bodies neither reflect nor absorb light; they are nonreactive to physical-energy light, and they so completely encircle and enshroud Havona as to hide it from the view of even near-by inhabited universes of time and space.

The great belt of dark gravity bodies is divided into two equal elliptical circuits by a unique space intrusion. The inner belt revolves counterclockwise; the outer revolves clockwise. These alternate directions of motion, coupled with the extraordinary mass of the dark bodies, so effectively equalize the lines of Havona gravity as to render the central universe a physically balanced and perfectly stabilized creation.

The inner procession of dark gravity bodies is tubular in arrangement, consisting of three circular groupings. A cross section of this circuit would exhibit three concentric circles of about equal density. The outer circuit of dark gravity bodies is arranged perpendicularly, being ten thousand times higher than the inner circuit. The up-and-down diameter of the outer circuit is fifty thousand times that of the transverse diameter.

The intervening space which exists between these two circuits of gravity bodies is *unique* in that nothing like it is to be found elsewhere in all the wide universe. This zone is characterized by enormous wave movements of an up-and-down nature and is permeated by tremendous energy activities of an unknown order.

In our opinion, nothing like the dark gravity bodies of the central universe will characterize the future evolution of the outer space levels; we regard these alternate processions of stupendous gravity-balancing bodies as unique in the master universe.

2. CONSTITUTION OF HAVONA

Spirit beings do not dwell in nebulous space; they do not inhabit ethereal worlds; they are domiciled on actual spheres of a material nature, worlds just as real as those on which mortals live. The Havona worlds are actual and literal, albeit their literal substance differs from the material organization of the planets of the seven superuniverses.

The physical realities of Havona represent an order of energy organization radically different from any prevailing in the evolutionary universes of space. Havona energies are threefold; superuniverse units of energy-matter contain a twofold energy charge, although one form of energy exists in negative and positive phases. The creation of the central universe is threefold (Trinity); the creation of a local universe (directly) is twofold, by a Creator Son and a Creative Spirit.

The material of Havona consists of the organization of exactly one thousand basic chemical elements and the balanced function of the seven forms of Havona energy. Each of these basic energies manifests seven phases of excitation, so that the Havona natives respond to forty-nine differing sensation stimuli. In other words, viewed from a purely physical standpoint, the natives of the central universe possess forty-nine specialized forms of sensation. The morontia senses are seventy, and the higher spiritual orders of reaction response vary in different types of beings from seventy to two hundred and ten.

None of the physical beings of the central universe would be visible to Urantians. Neither would any of the physical stimuli of those faraway worlds excite a reaction in your gross sense organs. If a Urantia mortal could be transported to Havona, he would there be deaf, blind, and utterly lacking in all other sense reactions; he could only function as a limited self-conscious being deprived of all environmental stimuli and all reactions thereto.

There are numerous physical phenomena and spiritual reactions transpiring in the central creation which are unknown on worlds such as Urantia. The basic organization of a threefold creation is wholly unlike that of the twofold constitution of the created universes of time and space.

All natural law is co-ordinated on a basis entirely different than in the dualenergy systems of the evolving creations. The entire central universe is organized in accordance with the threefold system of perfect and symmetrical control. Throughout the whole Paradise-Havona system there is maintained a perfect balance between all cosmic realities and all spiritual forces. Paradise, with an absolute grasp of material creation, perfectly regulates and maintains the physical energies of this central universe; the Eternal Son, as a part of his all-embracing spirit grasp, most perfectly sustains the spiritual status of all who indwell Havona. On Paradise nothing is experimental, and the Paradise-Havona system is a unit of creative perfection.

The universal spiritual gravity of the Eternal Son is amazingly active throughout the central universe. All spirit values and spiritual personalities are unceasingly drawn inward towards the abode of the Gods. This Godward urge is intense and inescapable. The ambition to attain God is stronger in the central universe, not because spirit gravity is stronger than in the outlying universes, but because those beings who have attained Havona are more fully spiritualized and hence more responsive to the ever-present action of the universal spirit-gravity pull of the Eternal Son.

Likewise does the Infinite Spirit draw all intellectual values Paradiseward. Throughout the central universe the mind gravity of the Infinite Spirit functions in liaison with the spirit gravity of the Eternal Son, and these together constitute the combined urge of the ascendant souls to find God, to attain Deity, to achieve Paradise, and to know the Father.

Havona is a spiritually perfect and physically stable universe. The control and balanced stability of the central universe appear to be perfect. Everything physical or spiritual is perfectly predictable, but mind phenomena and personality volition are not. We do infer that sin can be reckoned as impossible of occurrence, but we do this on the ground that the native freewill creatures of Havona have never been guilty of transgressing the will of Deity. Through all eternity these supernal beings have been consistently loyal to the Eternals of Days. Neither has sin appeared in any creature who has entered Havona as a pilgrim. There has never been an instance of misconduct by any creature of any group of personalities ever created in, or admitted to, the central Havona universe. So perfect and so divine are the methods and means of selection in the universes of time that never in the records of Havona has an error occurred; no mistakes have ever been made; no ascendant soul has ever been prematurely admitted to the central universe.

3. The Havona Worlds

Concerning the government of the central universe, there is none. Havona is so exquisitely perfect that no intellectual system of government is required. There are no regularly constituted courts, neither are there legislative assemblies; Havona requires only administrative direction. Here may be observed the height of the ideals of true *self*-government.

There is no need of government among such perfect and near-perfect intelligences. They stand in no need of regulation, for they are beings of native perfection interspersed with evolutionary creatures who have long since passed the scrutiny of the supreme tribunals of the superuniverses.

The administration of Havona is not automatic, but it is marvelously perfect and divinely efficient. It is chiefly planetary and is vested in the resident Eternal of Days, each Havona sphere being directed by one of these Trinity-origin personalities. Eternals of Days are not creators, but they are perfect administrators.

They teach with supreme skill and direct their planetary children with a perfection of wisdom bordering on absoluteness.

The billion spheres of the central universe constitute the training worlds of the high personalities native to Paradise and Havona and further serve as the final proving grounds for ascending creatures from the evolutionary worlds of time. In the execution of the Universal Father's great plan of creature ascension the pilgrims of time are landed on the receiving worlds of the outer or seventh circuit, and subsequent to increased training and enlarged experience, they are progressively advanced inward, planet by planet and circle by circle, until they finally attain the Deities and achieve residence on Paradise.

At present, although the spheres of the seven circuits are maintained in all their supernal glory, only about one per cent of all planetary capacity is utilized in the work of furthering the Father's universal plan of mortal ascension. About one tenth of one per cent of the area of these enormous worlds is dedicated to the life and activities of the Corps of the Finality, beings eternally settled in light and life who often sojourn and minister on the Havona worlds. These exalted beings have their personal residences on Paradise.

The planetary construction of the Havona spheres is entirely unlike that of the evolutionary worlds and systems of space. Nowhere else in all the grand universe is it convenient to utilize such enormous spheres as inhabited worlds. Triata physical constitution, coupled with the balancing effect of the immense dark gravity bodies, makes it possible so perfectly to equalize the physical forces and so exquisitely to balance the various attractions of this tremendous creation. Antigravity is also employed in the organization of the material functions and the spiritual activities of these enormous worlds.

The architecture, lighting, and heating, as well as the biologic and artistic embellishment, of the Havona spheres, are quite beyond the greatest possible stretch of human imagination. You cannot be told much about Havona; to understand its beauty and grandeur you must see it. But there are real rivers and lakes on these perfect worlds.

Spiritually these worlds are ideally appointed; they are fittingly adapted to their purpose of harboring the numerous orders of differing beings who function in the central universe. Manifold activities take place on these beautiful worlds which are far beyond human comprehension.

4. Creatures of the Central Universe

There are seven basic forms of living things and beings on the Havona worlds, and each of these basic forms exists in three distinct phases. Each of these three phases is divided into seventy major divisions, and each major division is composed of one thousand minor divisions, with yet other subdivisions, and so on. These basic life groups might be classified as:

1. Material.

2. Morontial.

3. Spiritual.

4. Absonite.

5. Ultimate.

6. Coabsolute.

7. Absolute.

Decay and death are not a part of the cycle of life on the Havona worlds. In the central universe the lower living things undergo the transmutation of materialization. They do change form and manifestation, but they do not resolve by process of decay and cellular death.

The Havona natives are all the offspring of the Paradise Trinity. They are without creature parents, and they are nonreproducing beings. We cannot portray the creation of these citizens of the central universe, beings who never were created. The entire story of the creation of Havona is an attempt to time-space an eternity fact which has no relation to time or space as mortal man comprehends them. But we must concede human philosophy a point of origin; even personalities far above the human level require a concept of "beginnings." Nevertheless, the Paradise-Havona system is eternal.

The natives of Havona live on the billion spheres of the central universe in the same sense that other orders of permanent citizenship dwell on their respective spheres of nativity. As the material order of sonship carries on the material, intellectual, and spiritual economy of a billion local systems in a superuniverse, so, in a larger sense, do the Havona natives live and function on the billion worlds of the central universe. You might possibly regard these Havoners as material creatures in the sense that the word "material" could be expanded to describe the physical realities of the divine universe.

There is a life that is native to Havona and possesses significance in and of itself. Havoners minister in many ways to Paradise descenders and to superuniverse ascenders, but they also live lives that are unique in the central universe and have relative meaning quite apart from either Paradise or the superuniverses.

As the worship of the faith sons of the evolutionary worlds ministers to the satisfaction of the Universal Father's love, so the exalted adoration of the Havona creatures satiates the perfect ideals of divine beauty and truth. As mortal man strives to do the will of God, these beings of the central universe live to gratify the ideals of the Paradise Trinity. In their very nature they *are* the will of God. Man rejoices in the goodness of God, Havoners exult in the divine beauty, while you both enjoy the ministry of the liberty of living truth.

Havoners have both optional present and future unrevealed destinies. And there is a progression of native creatures that is peculiar to the central universe, a progression that involves neither ascent to Paradise nor penetration of the superuniverses. This progression to higher Havona status may be suggested as follows:

1. Experiential progress outward from the first to the seventh circuit.

2. Progress inward from the seventh to the first circuit.

3. Intracircuit progress—progression within the worlds of a given circuit.

In addition to the Havona natives, the inhabitants of the central universe embrace numerous classes of pattern beings for various universe groups—advisers, directors, and teachers of their kind and to their kind throughout creation. All beings in all universes are fashioned along the lines of some one order of pattern creature living on some one of the billion worlds of Havona. Even the mortals of time have their goal and ideals of creature existence on the outer circuits of these pattern spheres on high.

Then there are those beings who have attained the Universal Father, and who are entitled to go and come, who are assigned here and there in the universes on missions of special service. And on every Havona world will be found the attainment candidates, those who have physically attained the central universe, but who have not yet achieved that spiritual development which will enable them to claim Paradise residence.

The Infinite Spirit is represented on the Havona worlds by a host of personalities, beings of grace and glory, who administer the details of the intricate intellectual and spiritual affairs of the central universe. On these worlds of divine perfection they perform the work indigenous to the normal conduct of this vast creation and, in addition, carry on the manifold tasks of teaching, training, and ministering to the enormous numbers of ascendant creatures who have climbed to glory from the dark worlds of space.

There are numerous groups of beings native to the Paradise-Havona system that are in no way directly associated with the ascension scheme of creature perfection attainment; therefore are they omitted from the personality classifications presented to the mortal races. Only the major groups of superhuman beings and those orders directly connected with your survival experience are herein presented.

Havona teems with the life of all phases of intelligent beings, who there seek to advance from lower to higher circuits in their efforts to attain higher levels of divinity realization and enlarged appreciation of supreme meanings, ultimate values, and absolute reality.

5. LIFE IN HAVONA

On Urantia you pass through a short and intense test during your initial life of material existence. On the mansion worlds and up through your system, constellation, and local universe, you traverse the morontia phases of ascension. On the training worlds of the superuniverse you pass through the true spirit stages of progression and are prepared for eventual transit to Havona. On the seven circuits of Havona your attainment is intellectual, spiritual, and experiential. And there is a definite task to be achieved on each of the worlds of each of these circuits.

Life on the divine worlds of the central universe is so rich and full, so complete and replete, that it wholly transcends the human concept of anything a created being could possibly experience. The social and economic activities of this eternal creation are entirely dissimilar to the occupations of material creatures living on evolutionary worlds like Urantia. Even the technique of Havona thought is unlike the process of thinking on Urantia.

The regulations of the central universe are fittingly and inherently natural; the rules of conduct are not arbitrary. In every requirement of Havona there is disclosed the reason of righteousness and the rule of justice. And these two factors, combined, equal what on Urantia would be denominated *fairness*. When you arrive in Havona, you will naturally enjoy doing things the way they should be done.

When intelligent beings first attain the central universe, they are received and domiciled on the pilot world of the seventh Havona circuit. As the new arrivals progress spiritually, attain identity comprehension of their superuniverse Master Spirit, they are transferred to the sixth circle. (It is from these arrangements in the central universe that the circles of progress in the human mind have been designated.) After ascenders have attained a realization of Supremacy and are thereby prepared for the Deity adventure, they are taken to the fifth circuit; and after attaining the Infinite Spirit, they are transferred to the fourth. Following the attainment of the Eternal Son, they are removed to the third; and when they have recognized the Universal Father, they go to sojourn on the second circuit of worlds, where they become more familiar with the Paradise hosts. Arrival on the first circuit of Havona signifies the acceptance of the candidates of time into the service of Paradise. Indefinitely, according to the length and nature of the creature ascension, they will tarry on the inner circuit of progressive spiritual attainment. From this inner circuit the ascending pilgrims pass inward to Paradise residence and admission to the Corps of the Finality.

During your sojourn in Havona as a pilgrim of ascent, you will be allowed to visit freely among the worlds of the circuit of your assignment. You will also be permitted to go back to the planets of those circuits you have previously traversed. And all this is possible to those who sojourn on the circles of Havona without the necessity of being ensupernaphimed. The pilgrims of time are able to equip themselves to traverse "achieved" space but must depend on the ordained technique to negotiate "unachieved" space; a pilgrim cannot leave Havona nor go forward beyond his assigned circuit without the aid of a transport supernaphim.

There is a refreshing originality about this vast central creation. Aside from the physical organization of matter and the fundamental constitution of the basic orders of intelligent beings and other living things, there is nothing in common between the worlds of Havona. Every one of these planets is an original, unique, and exclusive creation; each planet is a matchless, superb, and perfect production. And this diversity of individuality extends to all features of the physical, intellectual, and spiritual aspects of planetary existence. Each of these billion perfection spheres has been developed and embellished in accordance with the plans of the resident Eternal of Days. And this is just why no two of them are alike.

Not until you traverse the last of the Havona circuits and visit the last of the Havona worlds, will the tonic of adventure and the stimulus of curiosity disappear from your career. And then will the urge, the forward impulse of eternity, replace its forerunner, the adventure lure of time.

Monotony is indicative of immaturity of the creative imagination and inactivity of intellectual co-ordination with the spiritual endowment. By the time an ascendant mortal begins the exploration of these heavenly worlds, he has already attained emotional, intellectual, and social, if not spiritual, maturity.

Not only will you find undreamed-of changes confronting you as you advance from circuit to circuit in Havona, but your astonishment will be inexpressible as you progress from planet to planet within each circuit. Each of these billion study worlds is a veritable university of surprises. Continuing astonishment, unending wonder, is the experience of those who traverse these circuits and tour these gigantic spheres. Monotony is not a part of the Havona career.

Love of adventure, curiosity, and dread of monotony—these traits inherent in evolving human nature—were not put there just to aggravate and annoy you during your short sojourn on earth, but rather to suggest to you that death is only the beginning of an endless career of adventure, an everlasting life of anticipation, an eternal voyage of discovery.

Curiosity—the spirit of investigation, the urge of discovery, the drive of exploration—is a part of the inborn and divine endowment of evolutionary space creatures. These natural impulses were not given you merely to be frustrated and repressed. True, these ambitious urges must frequently be restrained during your short life on earth, disappointment must be often experienced, but they are to be fully realized and gloriously gratified during the long ages to come.

6. THE PURPOSE OF THE CENTRAL UNIVERSE

The range of the activities of seven-circuited Havona is enormous. In general, they may be described as:

1. Havonal.

2. Paradisiacal.

3. Ascendant-finite—Supreme-Ultimate evolutional.

Many superfinite activities take place in the Havona of the present universe age, involving untold diversities of absonite and other phases of mind and spirit functions. It is possible that the central universe serves many purposes which are not revealed to me, as it functions in numerous ways beyond the comprehension of the created mind. Nevertheless, I will endeavor to depict how this perfect creation ministers to the needs and contributes to the satisfactions of seven orders of universe intelligence.

1. *The Universal Father*—the First Source and Center. God the Father derives supreme parental satisfaction from the perfection of the central creation. He enjoys the experience of love satiety on near-equality levels. The perfect Creator is divinely pleased with the adoration of the perfect creature.

Havona affords the Father supreme achievement gratification. The perfection realization in Havona compensates for the time-space delay of the eternal urge of infinite expansion.

The Father enjoys the Havona reciprocation of the divine beauty. It satisfies the divine mind to afford a perfect pattern of exquisite harmony for all evolving universes.

Our Father beholds the central universe with perfect pleasure because it is a worthy revelation of spirit reality to all personalities of the universe of universes.

The God of universes has favorable regard for Havona and Paradise as the eternal power nucleus for all subsequent universe expansion in time and space.

The eternal Father views with never-ending satisfaction the Havona creation as the worthy and alluring goal for the ascension candidates of time, his mortal grandchildren of space achieving their Creator-Father's eternal home. And God takes pleasure in the Paradise-Havona universe as the eternal home of Deity and the divine family.

2. *The Eternal Son*—the Second Source and Center. To the Eternal Son the superb central creation affords eternal proof of the partnership effectiveness of the divine family—Father, Son, and Spirit. It is the spiritual and material basis for absolute confidence in the Universal Father.

Havona affords the Eternal Son an almost unlimited base for the ever-expanding realization of spirit power. The central universe afforded the Eternal Son the arena wherein he could safely and securely demonstrate the spirit and technique of the bestowal ministry for the instruction of his associate Paradise Sons.

Havona is the reality foundation for the Eternal Son's spirit-gravity control of the universe of universes. This universe affords the Son the gratification of parental craving, spiritual reproduction.

The Havona worlds and their perfect inhabitants are the first and the eternally final demonstration that the Son is the Word of the Father. Thereby is the consciousness of the Son as an infinite complement of the Father perfectly gratified.

And this universe affords the opportunity for the realization of reciprocation of equality fraternity between the Universal Father and the Eternal Son, and this constitutes the everlasting proof of the infinite personality of each.

3. *The Infinite Spirit*—the Third Source and Center. The Havona universe affords the Infinite Spirit proof of being the Conjoint Actor, the infinite representative of the unified Father-Son. In Havona the Infinite Spirit derives the combined satisfaction of functioning as a creative activity while enjoying the satisfaction of absolute coexistence with this divine achievement.

In Havona the Infinite Spirit found an arena wherein he could demonstrate the ability and willingness to serve as a potential mercy minister. In this perfect creation the Spirit rehearsed for the adventure of ministry in the evolutionary universes.

This perfect creation afforded the Infinite Spirit opportunity to participate in universe administration with both divine parents—to administer a universe as associate-Creator offspring, thereby preparing for the joint administration of the local universes as the Creative Spirit associates of the Creator Sons.

The Havona worlds are the mind laboratory of the creators of the cosmic mind and the ministers to every creature mind in existence. Mind is different on each Havona world and serves as the pattern for all spiritual and material creature intellects.

These perfect worlds are the mind graduate schools for all beings destined for Paradise society. They afforded the Spirit abundant opportunity to test out the technique of mind ministry on safe and advisory personalities.

Havona is a compensation to the Infinite Spirit for his widespread and unselfish work in the universes of space. Havona is the perfect home and retreat for the untiring Mind Minister of time and space.

4. *The Supreme Being*—the evolutionary unification of experiential Deity. The Havona creation is the eternal and perfect proof of the spiritual reality of the Supreme Being. This perfect creation is a revelation of the perfect and symmetrical spirit nature of God the Supreme before the beginnings of the power-personality synthesis of the finite reflections of the Paradise Deities in the experiential universes of time and space.

In Havona the power potentials of the Almighty are unified with the spiritual nature of the Supreme. This central creation is an exemplification of the future-eternal unity of the Supreme.

Havona is a perfect pattern of the universality potential of the Supreme. This universe is a finished portrayal of the future perfection of the Supreme and is suggestive of the potential of the Ultimate.

Havona exhibits finality of spirit values existing as living will creatures of supreme and perfect self-control; mind existing as ultimately equivalent to spirit; reality and unity of intelligence with an unlimited potential.

5. *The Co-ordinate Creator Sons.* Havona is the educational training ground where the Paradise Michaels are prepared for their subsequent adventures in universe creation. This divine and perfect creation is a pattern for every Creator Son. He strives to make his own universe eventually attain to these Paradise-Havona levels of perfection.

A Creator Son uses the creatures of Havona as personality-pattern possibilities for his own mortal children and spirit beings. The Michael and other Paradise Sons view Paradise and Havona as the divine destiny of the children of time.

The Creator Sons know that the central creation is the real source of that indispensable universe overcontrol which stabilizes and unifies their local universes. They know that the personal presence of the ever-present influence of the Supreme and of the Ultimate is in Havona.

Havona and Paradise are the source of a Michael Son's creative power. Here dwell the beings who co-operate with him in universe creation. From Paradise come the Universe Mother Spirits, the co-creators of local universes.

The Paradise Sons regard the central creation as the home of their divine parents—their home. It is the place they enjoy returning to ever and anon.

6. *The Co-ordinate Ministering Daughters.* The Universe Mother Spirits, cocreators of the local universes, secure their prepersonal training on the worlds of Havona in close association with the Spirits of the Circuits. In the central universe the Spirit Daughters of the local universes were duly trained in the methods of co-operation with the Sons of Paradise, all the while subject to the will of the Father.

On the worlds of Havona the Spirit and the Daughters of the Spirit find the mind patterns for all their groups of spiritual and material intelligences, and this central universe is the sometime destiny of those creatures which a Universe Mother Spirit jointly sponsors with an associated Creator Son.

The Universe Mother Creator remembers Paradise and Havona as the place of her origin and the home of the Infinite Mother Spirit, the abode of the personality presence of the Infinite Mind.

From this central universe also came the bestowal of the personal prerogatives of creatorship which a Universe Divine Minister employs as complemental to a Creator Son in the work of creating living will creatures.

And lastly, since these Daughter Spirits of the Infinite Mother Spirit will not likely ever return to their Paradise home, they derive great satisfaction from the universal reflectivity phenomenon associated with the Supreme Being in Havona and personalized in Majeston on Paradise.

7. *The Evolutionary Mortals of the Ascending Career.* Havona is the home of the pattern personality of every mortal type and the home of all superhuman personalities of mortal association who are not native to the creations of time.

These worlds provide the stimulus of all human impulses towards the attainment of true spirit values on the highest conceivable reality levels. Havona is the pre-Paradise training goal of every ascending mortal. Here mortals attain pre-Paradise Deity—the Supreme Being. Havona stands before every will creature as the portal to Paradise and God attainment.

Paradise is the home, and Havona the workshop and playground, of the finaliters. And every God-knowing mortal craves to be a finaliter.

The central universe is not only man's established destiny, but it is also the starting place of the eternal career of the finaliters as they shall sometime be started out on the undisclosed and universal adventure in the experience of exploring the infinity of the Universal Father.

Havona will unquestionably continue to function with absonite significance even in future universe ages which may witness space pilgrims attempting to find God on superfinite levels. Havona has capacity to serve as a training universe for absonite beings. It will probably be the finishing school when the seven superuniverses are functioning as the intermediate school for the graduates of the primary schools of outer space. And we incline to the opinion that the potentials of eternal Havona are really unlimited, that the central universe has eternal capacity to serve as an experiential training universe for all past, present, or future types of created beings.

[Presented by a Perfector of Wisdom commissioned thus to function by the Ancients of Days on Uversa.]

FIVE

THE SEVEN SUPERUNIVERSES

(PAPER 15)

A S FAR as the Universal Father is concerned—as a Father—the universes are virtually nonexistent; he deals with personalities; he is the Father of personalities. As far as the Eternal Son and the Infinite Spirit are concerned—as creator partners—the universes are localized and individual under the joint rule of the Creator Sons and the Creative Spirits. As far as the Paradise Trinity is concerned, outside Havona there are just seven inhabited universes, the seven superuniverses which hold jurisdiction over the circle of the first post-Havona space level. The Seven Master Spirits radiate their influence out from the central Isle, thus constituting the vast creation one gigantic wheel, the hub being the eternal Isle of Paradise, the seven spokes the radiations of the Seven Master Spirits, the rim the outer regions of the grand universe.

Early in the materialization of the universal creation the sevenfold scheme of the superuniverse organization and government was formulated. The first post-Havona creation was divided into seven stupendous segments, and the headquarters worlds of these superuniverse governments were designed and constructed. The present scheme of administration has existed from near eternity, and the rulers of these seven superuniverses are rightly called Ancients of Days.

Of the vast body of knowledge concerning the superuniverses, I can hope to tell you little, but there is operative throughout these realms a technique of intelligent control for both physical and spiritual forces, and the universal gravity presences there function in majestic power and perfect harmony. It is important first to gain an adequate idea of the physical constitution and material organization of the superuniverse domains, for then you will be the better prepared to grasp the significance of the marvelous organization provided for their spiritual government and for the intellectual advancement of the will creatures who dwell on the myriads of inhabited planets scattered hither and yon throughout these seven superuniverses.

1. THE SUPERUNIVERSE SPACE LEVEL

Within the limited range of the records, observations, and memories of the generations of a million or a billion of your short years, to all practical intents and purposes, Urantia and the universe to which it belongs are experiencing the adventure of one long and uncharted plunge into new space; but according to the records of Uversa, in accordance with older observations, in harmony with the more extensive experience and calculations of our order, and as a result of conclusions based on these and other findings, we know that the universes are engaged in an orderly, well-understood, and perfectly controlled processional, swinging in majestic grandeur around the First Great Source and Center and his residential universe.

We have long since discovered that the seven superuniverses traverse a great ellipse, a gigantic and elongated circle. Your solar system and other worlds of time are not plunging headlong, without chart and compass, into unmapped space. The local universe to which your system belongs is pursuing a definite and well-understood counterclockwise course around the vast swing that encircles the central universe. This cosmic path is well charted and is just as thoroughly known to the superuniverse star observers as the orbits of the planets constituting your solar system are known to Urantia astronomers.

Urantia is situated in a local universe and a superuniverse not fully organized, and your local universe is in immediate proximity to numerous partially completed physical creations. You belong to one of the relatively recent universes. But you are not, today, plunging on wildly into uncharted space nor swinging out blindly into unknown regions. You are following the orderly and predetermined path of the superuniverse space level. You are now passing through the very same space that your planetary system, or its predecessors, traversed ages ago; and some day in the remote future your system, or its successors, will again traverse the identical space through which you are now so swiftly plunging.

In this age and as direction is regarded on Urantia, superuniverse number one swings almost due north, approximately opposite, in an easterly direction, to the Paradise residence of the Great Sources and Centers and the central universe of Havona. This position, with the corresponding one to the west, represents the nearest physical approach of the spheres of time to the eternal Isle. Superuniverse number two is in the north, preparing for the westward swing, while number three now holds the northernmost segment of the great space path, having already turned into the bend leading to the southerly plunge. Number four is on the comparatively straightaway southerly flight, the advance regions now approaching opposition to the Great Centers. Number five has about left its position opposite the Center of Centers while continuing on the direct southerly course just preceding the eastward swing; number six occupies most of the southern curve, the segment from which your superuniverse has nearly passed.

Your local universe of Nebadon belongs to Orvonton, the seventh superuniverse, which swings on between superuniverses one and six, having not long since (as we reckon time) turned the southeastern bend of the superuniverse space level. Today, the solar system to which Urantia belongs is a few billion years past the swing around the southern curvature so that you are just now advancing beyond the southeastern bend and are moving swiftly through the long and comparatively straightaway northern path. For untold ages Orvonton will pursue this almost direct northerly course.

Urantia belongs to a system which is well out towards the borderland of your local universe; and your local universe is at present traversing the periphery of Orvonton. Beyond you there are still others, but you are far removed in space from those physical systems which swing around the great circle in comparative proximity to the Great Source and Center.

2. ORGANIZATION OF THE SUPERUNIVERSE

Only the Universal Father knows the location and actual number of inhabited worlds in space; he calls them all by name and number. I can give only the approximate number of inhabited or inhabitable planets, for some local universes have more worlds suitable for intelligent life than others. Nor have all projected local universes been organized. Therefore the estimates which I offer are solely for the purpose of affording some idea of the immensity of the material creation.

There are seven superuniverses in the grand universe, and they are constituted approximately as follows:

1. *The System.* The basic unit of the supergovernment consists of about one thousand inhabited or inhabitable worlds. Blazing suns, cold worlds, planets too near the hot suns, and other spheres not suitable for creature habitation are not included in this group. These one thousand worlds adapted to support life are called a system, but in the younger systems only a comparatively small number of these worlds may be inhabited. Each inhabited planet is presided over by a Planetary Prince, and each local system has an architectural sphere as its headquarters and is ruled by a System Sovereign.

2. *The Constellation.* One hundred systems (about 100,000 inhabitable planets) make up a constellation. Each constellation has an architectural headquarters sphere and is presided over by three Vorondadek Sons, the Most Highs. Each constellation also has a Faithful of Days in observation, an ambassador of the Paradise Trinity.

3. *The Local Universe.* One hundred constellations (about 10,000,000 inhabitable planets) constitute a local universe. Each local universe has a magnificent architectural headquarters world and is ruled by one of the co-ordinate Creator Sons of God of the order of Michael. Each universe is blessed by the presence of a Union of Days, a representative of the Paradise Trinity.

4. *The Minor Sector.* One hundred local universes (about 1,000,000,000 inhabitable planets) constitute a minor sector of the superuniverse government; it has a wonderful headquarters world, wherefrom its rulers, the Recents of Days, administer the affairs of the minor sector. There are three Recents of Days, Supreme Trinity Personalities, on each minor sector headquarters.

5. *The Major Sector.* One hundred minor sectors (about 100,000,000,000 inhabitable worlds) make one major sector. Each major sector is provided with a superb headquarters and is presided over by three Perfections of Days, Supreme Trinity Personalities.

6. *The Superuniverse.* Ten major sectors (about 1,000,000,000,000 inhabitable planets) constitute a superuniverse. Each superuniverse is provided with an enormous and glorious headquarters world and is ruled by three Ancients of Days.

7. *The Grand Universe.* Seven superuniverses make up the present organized grand universe, consisting of approximately seven trillion inhabitable worlds plus the architectural spheres and the one billion inhabited spheres of Havona. The superuniverses are ruled and administered indirectly and reflectively from Paradise by the Seven Master Spirits. The billion worlds of Havona are directly administered by the Eternals of Days, one such Supreme Trinity Personality presiding over each of these perfect spheres.

Excluding the Paradise-Havona spheres, the plan of universe organization provides for the following units:

Superuniverses 7
Major sectors 70
Minor sectors 7,000
Local universes 700,000
Constellations 70,000,000
Local systems 7,000,000,000
Inhabitable planets 7,000,000,000,000

Each of the seven superuniverses is constituted, approximately, as follows:

One system embraces, approximately . . . . 1,000 worlds
One constellation (100 systems) . . . . 100,000 worlds
One universe (100 constellations) . . . . 10,000,000 worlds
One minor sector (100 universes) . . . . 1,000,000,000 worlds
One major sector (100 minor sectors) . . . 100,000,000,000 worlds
One superuniverse (10 major sectors) . .1,000,000,000,000 worlds

All such estimates are approximations at best, for new systems are constantly evolving while other organizations are temporarily passing out of material existence.

3. THE SUPERUNIVERSE OF ORVONTON

Practically all of the starry realms visible to the naked eye on Urantia belong to the seventh section of the grand universe, the superuniverse of Orvonton. The vast Milky Way starry system represents the central nucleus of Orvonton, being largely beyond the borders of your local universe. This great aggregation of suns, dark islands of space, double stars, globular clusters, star clouds, spiral and other nebulae, together with myriads of individual planets, forms a watchlike, elongated-circular grouping of about one seventh of the inhabited evolutionary universes.

From the astronomical position of Urantia, as you look through the cross section of near-by systems to the great Milky Way, you observe that the spheres of Orvonton are traveling in a vast elongated plane, the breadth being far greater than the thickness and the length far greater than the breadth.

Observation of the so-called Milky Way discloses the comparative increase in Orvonton stellar density when the heavens are viewed in one direction, while on either side the density diminishes; the number of stars and other spheres decreases away from the chief plane of our material superuniverse. When the angle of observation is propitious, gazing through the main body of this realm of maximum density, you are looking toward the residential universe and the center of all things.

Of the ten major divisions of Orvonton, eight have been roughly identified by Urantian astronomers. The other two are difficult of separate recognition because you are obliged to view these phenomena from the inside. If you could look upon the superuniverse of Orvonton from a position far-distant in space, you would immediately recognize the ten major sectors of the seventh galaxy.

The rotational center of your minor sector is situated far away in the enormous and dense star cloud of Sagittarius, around which your local universe and its associated creations all move, and from opposite sides of the vast Sagittarius subgalactic system you may observe two great streams of star clouds emerging in stupendous stellar coils.

The nucleus of the physical system to which your sun and its associated planets belong is the center of the onetime Andronover nebula. This former spiral nebula was slightly distorted by the gravity disruptions associated with the events which were attendant upon the birth of your solar system, and which were occasioned by the near approach of a large neighboring nebula. This near collision changed Andronover into a somewhat globular aggregation but did not wholly destroy the two-way procession of the suns and their associated physical groups. Your solar system now occupies a fairly central position in one of the arms of this distorted spiral, situated about halfway from the center out towards the edge of the star stream.

The Sagittarius sector and all other sectors and divisions of Orvonton are in rotation around Uversa, and some of the confusion of Urantian star observers arises out of the illusions and relative distortions produced by the following multiple revolutionary movements:

1. The revolution of Urantia around its sun.

2. The circuit of your solar system about the nucleus of the former Andronover nebula.

3. The rotation of the Andronover stellar family and the associated clusters about the composite rotation-gravity center of the star cloud of Nebadon.

4. The swing of the local star cloud of Nebadon and its associated creations around the Sagittarius center of their minor sector.

5. The rotation of the one hundred minor sectors, including Sagittarius, about their major sector.

6. The whirl of the ten major sectors, the so-called star drifts, about the Uversa headquarters of Orvonton.

7. The movement of Orvonton and six associated superuniverses around Paradise and Havona, the counterclockwise processional of the superuniverse space level.

These multiple motions are of several orders: The space paths of your planet and your solar system are genetic, inherent in origin. The absolute counterclockwise motion of Orvonton is also genetic, inherent in the architectural plans of the master universe. But the intervening motions are of composite origin, being derived in part from the constitutive segmentation of matter-energy into the superuniverses and in part produced by the intelligent and purposeful action of the Paradise force organizers.

The local universes are in closer proximity as they approach Havona; the circuits are greater in number, and there is increased superimposition, layer upon layer. But farther out from the eternal center there are fewer and fewer systems, layers, circuits, and universes.

4. NEBULAE—THE ANCESTORS OF UNIVERSES

While creation and universe organization remain forever under the control of the infinite Creators and their associates, the whole phenomenon proceeds in accordance with an ordained technique and in conformity to the gravity laws of force, energy, and matter. But there is something of mystery associated with the universal force-charge of space; we quite understand the organization of the material creations from the ultimatonic stage forward, but we do not fully comprehend the cosmic ancestry of the ultimatons. We are confident that these ancestral forces have a Paradise origin because they forever swing through pervaded space in the exact gigantic outlines of Paradise. Though nonresponsive to Paradise gravity, this force-charge of space, the ancestor of all materialization, does always respond to the presence of nether Paradise, being apparently circuited in and out of the nether Paradise center.

The Paradise force organizers transmute space potency into primordial force and evolve this prematerial potential into the primary and secondary energy manifestations of physical reality. When this energy attains gravity-responding levels, the power directors and their associates of the superuniverse regime appear upon the scene and begin their never-ending manipulations designed to establish the manifold power circuits and energy channels of the universes of time and space. Thus does physical matter appear in space, and so is the stage set for the inauguration of universe organization.

This segmentation of energy is a phenomenon which has never been solved by the physicists of Nebadon. Their chief difficulty lies in the relative inaccessibility of the Paradise force organizers, for the living power directors, though they are competent to deal with space-energy, do not have the least conception of the origin of the energies they so skillfully and intelligently manipulate.

Paradise force organizers are nebulae originators; they are able to initiate about their space presence the tremendous cyclones of force which, when once started, can never be stopped or limited until the all-pervading forces are mobilized for the eventual appearance of the ultimatonic units of universe matter. Thus are brought into being the spiral and other nebulae, the mother wheels of the direct-origin suns and their varied systems. In outer space there may be seen ten different forms of nebulae, phases of primary universe evolution, and these vast energy wheels had the same origin as did those in the seven superuniverses.

Nebulae vary greatly in size and in the resulting number and aggregate mass of their stellar and planetary offspring. A sun-forming nebula just north of the borders of Orvonton, but within the superuniverse space level, has already given origin to approximately forty thousand suns, and the mother wheel is still throwing off suns, the majority of which are many times the size of yours. Some of the larger nebulae of outer space are giving origin to as many as one hundred million suns.

Nebulae are not directly related to any of the administrative units, such as minor sectors or local universes, although some local universes have been organized from the products of a single nebula. Each local universe embraces exactly one one-hundred-thousandth part of the total energy charge of a superuniverse irrespective of nebular relationship, for energy is not organized by nebulae—it is universally distributed.

Not all spiral nebulae are engaged in sun making. Some have retained control of many of their segregated stellar offspring, and their spiral appearance is occasioned by the fact that their suns pass out of the nebular arm in close formation but return by diverse routes, thus making it easy to observe them at one point but more difficult to see them when widely scattered on their different returning routes farther out and away from the arm of the nebula. There are not many sun-forming nebulae active in Orvonton at the present time, though Andromeda, which is outside the inhabited superuniverse, is very active. This far-distant nebula is visible to the naked eye, and when you view it, pause to consider that the light you behold left those distant suns almost one million years ago.

The Milky Way galaxy is composed of vast numbers of former spiral and other nebulae, and many still retain their original configuration. But as the result of internal catastrophes and external attraction, many have suffered such distortion and rearrangement as to cause these enormous aggregations to appear as gigantic luminous masses of blazing suns, like the Magellanic Cloud. The globular type of star clusters predominates near the outer margins of Orvonton.

The vast star clouds of Orvonton should be regarded as individual aggregations of matter comparable to the separate nebulae observable in the space regions external to the Milky Way galaxy. Many of the so-called star clouds of space, however, consist of gaseous material only. The energy potential of these stellar gas clouds is unbelievably enormous, and some of it is taken up by near-by suns and redispatched in space as solar emanations.

5. THE ORIGIN OF SPACE BODIES

The bulk of the mass contained in the suns and planets of a superuniverse originates in the nebular wheels; very little of superuniverse mass is organized by the direct action of the power directors (as in the construction of architectural spheres), although a constantly varying quantity of matter originates in open space.

As to origin, the majority of the suns, planets, and other spheres can be classified in one of the following ten groups:

1. *Concentric Contraction Rings.* Not all nebulae are spiral. Many an immense nebula, instead of splitting into a double star system or evolving as a spiral, undergoes condensation by multiple-ring formation. For long periods such a nebula appears as an enormous central sun surrounded by numerous gigantic clouds of encircling, ring-appearing formations of matter.

2. *The Whirled Stars* embrace those suns which are thrown off the great mother wheels of highly heated gases. They are not thrown off as rings but in right- and left-handed processions. Whirled stars are also of origin in other-than-spiral nebulae.

3. *Gravity-explosion Planets.* When a sun is born of a spiral or of a barred nebula, not infrequently it is thrown out a considerable distance. Such a sun is highly gaseous, and subsequently, after it has somewhat cooled and condensed, it may chance to swing near some enormous mass of matter, a gigantic sun or a dark island of space. Such an approach may not be near enough to result in collision but still near enough to allow the gravity pull of the greater

body to start tidal convulsions in the lesser, thus initiating a series of tidal upheavals which occur simultaneously on opposite sides of the convulsed sun. At their height these explosive eruptions produce a series of varying-sized aggregations of matter which may be projected beyond the gravity-reclamation zone of the erupting sun, thus becoming stabilized in orbits of their own around one of the two bodies concerned in this episode. Later on the larger collections of matter unite and gradually draw the smaller bodies to themselves. In this way many of the solid planets of the lesser systems are brought into existence. Your own solar system had just such an origin.

4. *Centrifugal Planetary Daughters.* Enormous suns, when in certain stages of development, and if their revolutionary rate greatly accelerates, begin to throw off large quantities of matter which may subsequently be assembled to form small worlds that continue to encircle the parent sun.

5. *Gravity-deficiency Spheres.* There is a critical limit to the size of individual stars. When a sun reaches this limit, unless it slows down in revolutionary rate, it is doomed to split; sun fission occurs, and a new double star of this variety is born. Numerous small planets may be subsequently formed as a by-product of this gigantic disruption.

6. *Contractural Stars.* In the smaller systems the largest outer planet sometimes draws to itself its neighboring worlds, while those planets near the sun begin their terminal plunge. With your solar system, such an end would mean that the four inner planets would be claimed by the sun, while the major planet, Jupiter, would be greatly enlarged by capturing the remaining worlds. Such an end of a solar system would result in the production of two adjacent but unequal suns, one type of double star formation. Such catastrophes are infrequent except out on the fringe of the superuniverse starry aggregations.

7. *Cumulative Spheres.* From the vast quantity of matter circulating in space, small planets may slowly accumulate. They grow by meteoric accretion and by minor collisions. In certain sectors of space, conditions favor such forms of planetary birth. Many an inhabited world has had such an origin.

Some of the dense dark islands are the direct result of the accretions of transmuting energy in space. Another group of these dark islands have come into being by the accumulation of enormous quantities of cold matter, mere fragments and meteors, circulating through space. Such aggregations of matter have never been hot and, except for density, are in composition very similar to Urantia.

8. *Burned-out Suns.* Some of the dark islands of space are burned-out isolated suns, all available space-energy having been emitted. The organized units of matter approximate full condensation, virtual complete consolidation; and it requires ages upon ages for such enormous masses of highly condensed matter to be recharged in the circuits of space and thus to be prepared for new cycles of universe function following a collision or some equally revivifying cosmic happening.

9. *Collisional Spheres.* In those regions of thicker clustering, collisions are not uncommon. Such an astronomic readjustment is accompanied by tremendous energy changes and matter transmutations. Collisions involving dead suns are peculiarly influential in creating widespread energy fluctuations. Collisional debris often constitutes the material nucleuses for the subsequent formation of planetary bodies adapted to mortal habitation.

10. *Architectural Worlds.* These are the worlds which are built according to plans and specifications for some special purpose, such as Salvington, the headquarters of your local universe, and Uversa, the seat of government of our superuniverse.

There are numerous other techniques for evolving suns and segregating planets, but the foregoing procedures suggest the methods whereby the vast majority of stellar systems and planetary families are brought into existence. To undertake to describe all the various techniques involved in stellar metamorphosis and planetary evolution would require the narration of almost one hundred different modes of sun formation and planetary origin. As your star students scan the heavens, they will observe phenomena indicative of all these modes of stellar evolution, but they will seldom detect evidence of the formation of those small, nonluminous collections of matter which serve as inhabited planets, the most important of the vast material creations.

6. THE SPHERES OF SPACE

Irrespective of origin, the various spheres of space are classifiable into the following major divisions:

1. The suns—the stars of space.
2. The dark islands of space.
3. Minor space bodies—comets, meteors, and planetesimals.
4. The planets, including the inhabited worlds.
5. Architectural spheres—worlds made to order.

With the exception of the architectural spheres, all space bodies have had an evolutionary origin, evolutionary in the sense that they have not been brought into being by fiat of Deity, evolutionary in the sense that the creative acts of God have unfolded by a time-space technique through the operation of many of the created and eventuated intelligences of Deity.

The Suns. These are the stars of space in all their various stages of existence. Some are solitary evolving space systems; others are double stars, contracting or disappearing planetary systems. The stars of space exist in no less than a thousand different states and stages. You are familiar with suns that emit light accompanied by heat; but there are also suns which shine without heat.

The trillions upon trillions of years that an ordinary sun will continue to give out heat and light well illustrates the vast store of energy which each unit of matter contains. The actual energy stored in these invisible particles of physical matter is well-nigh unimaginable. And this energy becomes almost wholly available as light when subjected to the tremendous heat pressure and the associated energy activities which prevail in the interior of the blazing suns. Still other conditions enable these suns to transform and send forth much of the energy of space which comes their way in the established space circuits. Many phases of physical energy and all forms of matter are attracted to, and subsequently distributed by, the solar dynamos. In this way the suns serve as local accelerators of energy circulation, acting as automatic power-control stations.

The superuniverse of Orvonton is illuminated and warmed by more than ten trillion blazing suns. These suns are the stars of your observable astronomic system. More than two trillion are too distant and too small ever to be seen from Urantia. But in the master universe there are as many suns as there are glasses of water in the oceans of your world.

The Dark Islands of Space. These are the dead suns and other large aggregations of matter devoid of light and heat. The dark islands are sometimes enormous in mass and exert a powerful influence in universe equilibrium and energy manipulation. The density of some of these large masses is well-nigh unbelievable. And this great concentration of mass enables these dark islands to function as powerful balance wheels, holding large neighboring systems in effective leash. They hold the gravity balance of power in many constellations; many physical systems which would otherwise speedily dive to destruction in near-by suns are held securely in the gravity grasp of these guardian dark islands. It is because of this function that we can locate them accurately. We have measured the gravity pull of the luminous bodies, and we can therefore calculate the exact size and location of the dark islands of space which so effectively function to hold a given system steady in its course.

Minor Space Bodies. The meteors and other small particles of matter circulating and evolving in space constitute an enormous aggregate of energy and material substance.

Many comets are unestablished wild offspring of the solar mother wheels, which are being gradually brought under control of the central governing sun. Comets also have numerous other origins. A comet's tail points away from the attracting body or sun because of the electrical reaction of its highly expanded gases and because of the actual pressure of light and other energies emanating from the sun. This phenomenon constitutes one of the positive proofs of the reality of light and its associated energies; it demonstrates that light has weight. Light is a real substance, not simply waves of hypothetical ether.

The Planets. These are the larger aggregations of matter which follow an orbit around a sun or some other space body; they range in size from planetesimals to enormous gaseous, liquid, or solid spheres. The cold worlds which have been built up by the assemblage of floating space material, when they happen to be in proper relation to a near-by sun, are the more ideal planets to harbor intelligent inhabitants. The dead suns are not, as a rule, suited to life; they are usually too far away from a living, blazing sun, and further, they are altogether too massive; gravity is tremendous at the surface.

In your superuniverse not one cool planet in forty is habitable by beings of your order. And, of course, the superheated suns and the frigid outlying worlds are unfit to harbor higher life. In your solar system only three planets are at present suited to harbor life. Urantia, in size, density, and location, is in many respects ideal for human habitation.

The laws of physical-energy behavior are basically universal, but local influences have much to do with the physical conditions which prevail on individual planets and in local systems. An almost endless variety of creature life and other living manifestations characterizes the countless worlds of space. There are, however, certain points of similarity in a group of worlds associated in a given system, while there also is a universe pattern of intelligent life. There are physical relationships among those planetary systems which belong to the same physical circuit, and which closely follow each other in the endless swing around the circle of universes.

7. The Architectural Spheres

While each superuniverse government presides near the center of the evolutionary universes of its space segment, it occupies a world made to order and is peopled by accredited personalities. These headquarters worlds are architectural spheres, space bodies specifically constructed for their special purpose. While sharing the light of near-by suns, these spheres are independently lighted and heated. Each has a sun which gives forth light without heat, like the satellites of Paradise, while each is supplied with heat by the circulation of certain energy currents near the surface of the sphere. These headquarters worlds belong to one of the greater systems situated near the astronomical center of their respective superuniverses.

Time is standardized on the headquarters of the superuniverses. The standard day of the superuniverse of Orvonton is equal to almost thirty days of Urantia time, and the Orvonton year equals one hundred standard days. This Uversa year is standard in the seventh superuniverse, and it is twenty-two minutes short of three thousand days of Urantia time, about eight and one fifth of your years.

The headquarters worlds of the seven superuniverses partake of the nature and grandeur of Paradise, their central pattern of perfection. In reality, all headquarters worlds are paradisiacal. They are indeed heavenly abodes, and they increase in material size, morontia beauty, and spirit glory from Jerusem to the central Isle. And all the satellites of these headquarters worlds are also architectural spheres.

The various headquarters worlds are provided with every phase of material and spiritual creation. All kinds of material, morontial, and spiritual beings are at home on these rendezvous worlds of the universes. As mortal creatures ascend the universe, passing from the material to the spiritual realms, they never lose their appreciation for, and enjoyment of, their former levels of existence.

Jerusem, the headquarters of your local system of Satania, has its seven worlds of transition culture, each of which is encircled by seven satellites, among which are the seven mansion worlds of morontia detention, man's first postmortal residence. As the term heaven has been used on Urantia, it has sometimes meant these seven mansion worlds, the first mansion world being denominated the first heaven, and so on to the seventh.

Edentia, the headquarters of your constellation of Norlatiadek, has its seventy satellites of socializing culture and training, on which ascenders sojourn upon the completion of the Jerusem regime of personality mobilization, unification, and realization.

Salvington, the capital of Nebadon, your local universe, is surrounded by ten university clusters of forty-nine spheres each. Hereon is man spiritualized following his constellation socialization.

Uminor the third, the headquarters of your minor sector, Ensa, is surrounded by the seven spheres of the higher physical studies of the ascendant life.

Umajor the fifth, the headquarters of your major sector, Splandon, is surrounded by the seventy spheres of the advancing intellectual training of the superuniverse.

Uversa, the headquarters of Orvonton, your superuniverse, is immediately surrounded by the seven higher universities of advanced spiritual training for ascending will creatures. Each of these seven clusters of wonder spheres consists of seventy specialized worlds containing thousands upon thousands of replete institutions and organizations devoted to universe training and spirit culture wherein the pilgrims of time are re-educated and re-examined preparatory to their long flight to Havona. The arriving pilgrims of time are always received on these associated worlds, but the departing graduates are always dispatched for Havona direct from the shores of Uversa.

Uversa is the spiritual and administrative headquarters for approximately one trillion inhabited or inhabitable worlds. The glory, grandeur, and perfection of the Orvonton capital surpass any of the wonders of the time-space creations.

If all the projected local universes and their component parts were established, there would be slightly less than five hundred billion architectural worlds in the seven superuniverses.

8. ENERGY CONTROL AND REGULATION

The headquarters spheres of the superuniverses are so constructed that they are able to function as efficient power-energy regulators for their various sectors, serving as focal points for the directionization of energy to their component local universes. They exert a powerful influence over the balance and control of the physical energies circulating through organized space.

Further regulative functions are performed by the superuniverse power centers and physical controllers, living and semiliving intelligent entities constituted for this express purpose. These power centers and controllers are difficult of understanding; the lower orders are not volitional, they do not possess will, they do not choose, their functions are very intelligent but apparently automatic and inherent in their highly specialized organization. The power centers and physical controllers of the superuniverses assume direction and partial control of the thirty energy systems which comprise the gravita domain. The physical-energy circuits administered by the power centers of Uversa require a little over 968 million years to complete the encirclement of the superuniverse.

Evolving energy has substance; it has weight, although weight is always relative, depending on revolutionary velocity, mass, and antigravity. Mass in matter tends to retard velocity in energy; and the anywhere-present velocity of energy represents: the initial endowment of velocity, minus retardation by mass encountered in transit, plus the regulatory function of the living energy controllers of the superuniverse and the physical influence of near-by highly heated or heavily charged bodies.

The universal plan for the maintenance of equilibrium between matter and energy necessitates the everlasting making and unmaking of the lesser material units. The Universe Power Directors have the ability to condense and detain, or to expand and liberate, varying quantities of energy.

Given a sufficient duration of retarding influence, gravity would eventually convert all energy into matter were it not for two factors: First, because of the antigravity influences of the energy controllers, and second, because organized matter tends to disintegrate under certain conditions found in very hot stars and under certain peculiar conditions in space near highly energized cold bodies of condensed matter.

When mass becomes overaggregated and threatens to unbalance energy, to deplete the physical power circuits, the physical controllers intervene unless gravity's own further tendency to over-materialize energy is defeated by the occurrence of a collision among the dead giants of space, thus in an instant completely dissipating the cumulative collections of gravity. In these collisional episodes enormous masses of matter are suddenly converted into the rarest form of energy, and the struggle for universal equilibrium is begun anew. Eventually the larger physical systems become stabilized, become physically settled, and are swung into the balanced and established circuits of the superuniverses. Subsequent to this event no more collisions or other devastating catastrophes will occur in such established systems.

During the times of plus energy there are power disturbances and heat fluctuations accompanied by electrical manifestations. During times of minus energy there are increased tendencies for matter to aggregate, condense, and to get out of control in the more delicately balanced circuits, with resultant tidal or collisional adjustments which quickly restore the balance between circulating energy and more literally stabilized matter. To forecast and otherwise to understand such likely behavior of the blazing suns and the dark islands of space is one of the tasks of the celestial star observers.

We are able to recognize most of the laws governing universe equilibrium and to predict much pertaining to universe stability. Practically, our forecasts are reliable, but we are always confronted by certain forces which are not wholly amenable to the laws of energy control and matter behavior known to us. The predictability of all physical phenomena becomes increasingly difficult as we proceed outward in the universes from Paradise. As we pass beyond the borders of the personal administration of the Paradise Rulers, we are confronted with increasing inability to reckon in accordance with the standards established and the experience acquired in connection with observations having exclusively to do with the physical phenomena of the near-by astronomic systems. Even in the realms of the seven superuniverses we are living in the midst of force actions and energy reactions which pervade all our domains and extend in unified equilibrium on through all regions of outer space.

The farther out we go, the more certainly we encounter those variational and unpredictable phenomena which are so unerringly characteristic of the unfathomable presence-performances of the Absolutes and the experiential Deities. And these phenomena must be indicative of some universal overcontrol of all things.

The superuniverse of Orvonton is apparently now running down; the outer universes seem to be winding up for unparalleled future activities; the central Havona universe is eternally stabilized. Gravity and absence of heat (cold) organize and hold matter together; heat and antigravity disrupt matter and dissipate energy. The living power directors and force organizers are the secret of the special control and intelligent direction of the endless metamorphoses of universe making, unmaking, and remaking. Nebulae may disperse, suns burn out, systems vanish, and planets perish, but the universes do not run down.

9. CIRCUITS OF THE SUPERUNIVERSES

The universal circuits of Paradise do actually pervade the realms of the seven superuniverses. These presence circuits are: the personality gravity of the Universal Father, the spiritual gravity of the Eternal Son, the mind gravity of the Conjoint Actor, and the material gravity of the eternal Isle.

In addition to the universal Paradise circuits and in addition to the presence-performances of the Absolutes and the experiential

Deities, there function within the superuniverse space level only two energy-circuit divisions or power segregations: the superuniverse circuits and the local universe circuits.

The Superuniverse Circuits:

1. The unifying intelligence circuit of one of the Seven Master Spirits of Paradise. Such a cosmic-mind circuit is limited to a single superuniverse.

2. The reflective-service circuit of the seven Reflective Spirits in each superuniverse.

3. The secret circuits of the Mystery Monitors, in some manner interassociated and routed by Divinington to the Universal Father on Paradise.

4. The circuit of the intercommunion of the Eternal Son with his Paradise Sons.

5. The flash presence of the Infinite Spirit.

6. The broadcasts of Paradise, the space reports of Havona.

7. The energy circuits of the power centers and the physical controllers.

The Local Universe Circuits:

1. The bestowal spirit of the Paradise Sons, the Comforter of the bestowal worlds. The Spirit of Truth, the spirit of Michael on Urantia.

2. The circuit of the Divine Ministers, the local universe Mother Spirits, the Holy Spirit of your world.

3. The intelligence-ministry circuit of a local universe, including the diversely functioning presence of the adjutant mind-spirits.

When there develops such a spiritual harmony in a local universe that its individual and combined circuits become indistinguishable from those of the superuniverse, when such identity of function and oneness of ministry actually prevail, then does the local universe immediately swing into the settled circuits of light and life, becoming at once eligible for admission into the spiritual confederation of the perfected union of the supercreation. The requisites for admission to the councils of the Ancients of Days, membership in the superuniverse confederation, are:

1. *Physical Stability.* The stars and planets of a local universe must be in equilibrium; the periods of immediate stellar metamorphosis must be over. The universe must be proceeding on a clear track; its orbit must be safely and finally settled.

2. *Spiritual Loyalty.* There must exist a state of universal recognition of, and loyalty to, the Sovereign Son of God who presides over the affairs of such a local universe. There must have come into being a state of harmonious cooperation between the individual planets, systems, and constellations of the entire local universe.

Your local universe is not even reckoned as belonging to the settled physical order of the superuniverse, much less as holding membership in the recognized spiritual family of the supergovernment. Although Nebadon does not yet have representation on Uversa, we of the superuniverse government are dispatched to its worlds on special missions from time to time, even as I have come to Urantia directly from Uversa. We lend every possible assistance to your directors and rulers in the solution of their difficult problems; we are desirous of seeing your universe qualified for full admission into the associated creations of the superuniverse family.

10. Rulers of the Superuniverses

The headquarters of the superuniverses are the seats of the high spiritual government of the time-space domains. The executive branch of the supergovernment, taking origin in the Councils of the Trinity, is immediately directed by one of the Seven Master Spirits of supreme supervision, beings who sit upon seats of Paradise authority and administer the superuniverses through the Seven Supreme Executives stationed on the seven special worlds of the Infinite Spirit, the outermost satellites of Paradise.

The superuniverse headquarters are the abiding places of the Reflective Spirits and the Reflective Image Aids. From this midway position these marvelous beings conduct their tremendous reflectivity operations, thus ministering to the central universe above and to the local universes below.

Each superuniverse is presided over by three Ancients of Days, the joint chief executives of the supergovernment. In its executive branch the personnel of the superuniverse government consists of seven different groups:

1. Ancients of Days.
2. Perfectors of Wisdom.
3. Divine Counselors.
4. Universal Censors.
5. Mighty Messengers.
6. Those High in Authority.
7. Those without Name and Number.

The three Ancients of Days are immediately assisted by a corps of one billion Perfectors of Wisdom, with whom are associated three billion Divine Counselors. One billion Universal Censors are attached to each superuniverse administration. These three groups are Co-ordinate Trinity Personalities, taking origin directly and divinely in the Paradise Trinity.

The remaining three orders, Mighty Messengers, Those High in Authority, and Those without Name and Number, are glorified ascendant mortals. The first of these orders came up through the ascendant regime and passed through Havona in the days of Grandfanda. Having attained Paradise, they were mustered into the Corps of the Finality, embraced by the Paradise Trinity, and subsequently assigned to the supernal service of the Ancients of Days. As a class, these three orders are known as Trinitized Sons of Attainment, being of dual origin but now of Trinity service. Thus was the executive branch of the superuniverse government enlarged to include the glorified and perfected children of the evolutionary worlds.

The co-ordinate council of the superuniverse is composed of the seven executive groups previously named and the following sector rulers and other regional overseers:

1. Perfections of Days—the rulers of the superuniverse major sectors.

2. Recents of Days—the directors of the superuniverse minor sectors.

3. Unions of Days—the Paradise advisers to the rulers of the local universes.

4. Faithfuls of Days—the Paradise counselors to the Most High rulers of the constellation governments.

5. Trinity Teacher Sons who may chance to be on duty at superuniverse headquarters.

6. Eternals of Days who may happen to be present at superuniverse headquarters.

7. The seven Reflective Image Aids—the spokesmen of the seven Reflective Spirits and through them representatives of the Seven Master Spirits of Paradise.

The Reflective Image Aids also function as the representatives of numerous groups of beings who are influential in the superuniverse governments, but who are not, at present, for various reasons, fully active in their individual capacities. Embraced within this group are: the evolving superuniverse personality manifestation of the Supreme Being, the Unqualified Supervisors of the Supreme, the Qualified Vicegerents of the Ultimate, the unnamed liaison reflectivators of Majeston, and the superpersonal spirit representatives of the Eternal Son.

At almost all times it is possible to find representatives of all groups of created beings on the headquarters worlds of the superuniverses. The routine ministering work of the superuniverses is performed by the mighty seconaphim and by other members of the vast family of the Infinite Spirit. In the work of these marvelous centers of superuniverse administration, control, ministry, and executive judgment, the intelligences of every sphere of universal life are mingled in effective service, wise administration, loving ministry, and just judgment.

The superuniverses do not maintain any sort of ambassadorial representation; they are completely isolated from each other. They know of mutual affairs only through the Paradise clearinghouse maintained by the Seven Master Spirits. Their rulers work in the councils of divine wisdom for the welfare of their own superuniverses regardless of what may be transpiring in other sections of the universal creation. This isolation of the superuniverses will persist until such time as their co-ordination is achieved by the more complete factualization of the personality-sovereignty of the evolving experiential Supreme Being.

11. THE DELIBERATIVE ASSEMBLY

It is on such worlds as Uversa that the beings representative of the autocracy of perfection and the democracy of evolution meet face to face. The executive branch of the supergovernment origi-

nates in the realms of perfection; the legislative branch springs from
the flowering of the evolutionary universes.

The deliberative assembly of the superuniverse is confined to
the headquarters world. This legislative or advisory council consists
of seven houses, to each of which every local universe admitted to
the superuniverse councils elects a native representative. These representatives are chosen by the high councils of such local universes
from among the ascending-pilgrim graduates of Orvonton who are
tarrying on Uversa, accredited for transport to Havona. The average
term of service is about one hundred years of superuniverse standard time.

Never have I known of a disagreement between the Orvonton
executives and the Uversa assembly. Never yet, in the history of our
superuniverse, has the deliberative body ever passed a recommen-
dation that the executive division of the supergovernment has even
hesitated to carry out. There always has prevailed the most perfect
harmony and working agreement, all of which testifies to the fact
that evolutionary beings can really attain the heights of perfected
wisdom which qualifies them to consort with the personalities of
perfect origin and divine nature. The presence of the deliberative
assemblies on the superuniverse headquarters reveals the wisdom,
and foreshadows the ultimate triumph, of the whole vast evolutionary concept of the Universal Father and his Eternal Son.

12. THE SUPREME TRIBUNALS

When we speak of executive and deliberative branches of the
Uversa government, you may, from the analogy of certain forms of
Urantian civil government, reason that we must have a third or judicial branch, and we do; but it does not have a separate personnel.
Our courts are constituted as follows: There presides, in accordance with the nature and gravity of the case, an Ancient of Days,
a Perfector of Wisdom, or a Divine Counselor. The evidence for or
against an individual, a planet, system, constellation, or universe is
presented and interpreted by the Censors. The defense of the children of time and the evolutionary planets is offered by the Mighty
Messengers, the official observers of the superuniverse government
to the local universes and systems. The attitude of the higher government is portrayed by Those High in Authority. And ordinarily
the verdict is formulated by a varying-sized commission consisting

equally of Those without Name and Number and a group of understanding personalities chosen from the deliberative assembly.

The courts of the Ancients of Days are the high review tribunals for the spiritual adjudication of all component universes. The Sovereign Sons of the local universes are supreme in their own domains; they are subject to the supergovernment only in so far as they voluntarily submit matters for counsel or adjudication by the Ancients of Days except in matters involving the extinction of will creatures. Mandates of judgment originate in the local universes, but sentences involving the extinction of will creatures are always formulated on, and executed from, the headquarters of the superuniverse. The Sons of the local universes can decree the survival of mortal man, but only the Ancients of Days may sit in executive judgment on the issues of eternal life and death.

In all matters not requiring trial, the submission of evidence, the Ancients of Days or their associates render decisions, and these rulings are always unanimous. We are here dealing with the councils of perfection. There are no disagreements nor minority opinions in the decrees of these supreme and superlative tribunals.

With certain few exceptions the supergovernments exercise jurisdiction over all things and all beings in their respective domains. There is no appeal from the rulings and decisions of the superuniverse authorities since they represent the concurred opinions of the Ancients of Days and that Master Spirit who, from Paradise, presides over the destiny of the superuniverse concerned.

13. The Sector Governments

A *major sector* comprises about one tenth of a superuniverse and consists of one hundred minor sectors, ten thousand local universes, about one hundred billion inhabitable worlds. These major sectors are administered by three Perfections of Days, Supreme Trinity Personalities.

The courts of the Perfections of Days are constituted much as are those of the Ancients of Days except that they do not sit in spiritual judgment upon the realms. The work of these major sector governments has chiefly to do with the intellectual status of a far-flung creation. The major sectors detain, adjudicate, dispense, and tabulate, for reporting to the courts of the Ancients of Days, all matters of superuniverse importance of a routine and administrative nature which are not immediately concerned with the spiritual administration of the realms or with the outworking of the mortal-ascension plans of the Paradise Rulers. The personnel of a major sector government is no different from that of the superuniverse.

As the magnificent satellites of Uversa are concerned with your final spiritual preparation for Havona, so are the seventy satellites of Umajor the fifth devoted to your superuniverse intellectual training and development. From all Orvonton, here are gathered together the wise beings who labor untiringly to prepare the mortals of time for their further progress towards the career of eternity. Most of this training of ascending mortals is conducted on the seventy study worlds.

The *minor sector* governments are presided over by three Re-cents of Days. Their administration is concerned mainly with the physical control, unification, stabilization, and routine co-ordina-tion of the administration of the component local universes. Each minor sector embraces as many as one hundred local universes, ten thousand constellations, one million systems, or about one billion inhabitable worlds.

Minor sector headquarters worlds are the grand rendezvous of the Master Physical Controllers. These headquarters worlds are surrounded by the seven instruction spheres which constitute the entrance schools of the superuniverse and are the centers of training for physical and administrative knowledge concerning the universe of universes.

The administrators of the minor sector governments are under the immediate jurisdiction of the major sector rulers. The Recents of Days receive all reports of observations and co-ordinate all rec-ommendations which come up to a superuniverse from the Unions of Days who are stationed as Trinity observers and advisers on the headquarters spheres of the local universes and from the Faithfuls of Days who are similarly attached to the councils of the Most Highs at the headquarters of the constellations. All such reports are transmit-ted to the Perfections of Days on the major sectors, subsequently to be passed on to the courts of the Ancients of Days. Thus the Trinity regime extends from the constellations of the local universes up to the headquarters of the superuniverse. The local system headquar-ters do not have Trinity representatives.

14. PURPOSES OF THE SEVEN SUPERUNIVERSES

There are seven major purposes which are being unfolded in the evolution of the seven superuniverses. Each major purpose in superuniverse evolution will find fullest expression in only one of the seven superuniverses, and therefore does each superuniverse have a special function and a unique nature.

Orvonton, the seventh superuniverse, the one to which your local universe belongs, is known chiefly because of its tremendous and lavish bestowal of merciful ministry to the mortals of the realms. It is renowned for the manner in which justice prevails as tempered by mercy and power rules as conditioned by patience, while the sacrifices of time are freely made to secure the stabilization of eternity. Orvonton is a universe demonstration of love and mercy.

It is, however, very difficult to describe our conception of the true nature of the evolutionary purpose which is unfolding in Orvonton, but it may be suggested by saying that in this supercreation we feel that the six unique purposes of cosmic evolution as manifested in the six associated supercreations are here being interassociated into a meaning-of-the-whole; and it is for this reason that we have sometimes conjectured that the evolved and finished personalization of God the Supreme will in the remote future and from Uversa rule the perfected seven superuniverses in all the experiential majesty of his then attained almighty sovereign power.

As Orvonton is unique in nature and individual in destiny, so also is each of its six associated superuniverses. A great deal that is going on in Orvonton is not, however, revealed to you, and of these unrevealed features of Orvonton life, many are to find most complete expression in some other superuniverse. The seven purposes of superuniverse evolution are operative throughout all seven superuniverses, but each supercreation will give fullest expression to only one of these purposes. To understand more about these superuniverse purposes, much that you do not understand would have to be revealed, and even then you would comprehend but little. This entire narrative presents only a fleeting glimpse of the immense creation of which your world and local system are a part.

Your world is called Urantia, and it is number 606 in the planetary group, or system, of Satania. This system has at present 619 inhabited worlds, and more than two hundred additional planets are evolving favorably toward becoming inhabited worlds at some future time.

Satania has a headquarters world called Jerusem, and it is system number twenty-four in the constellation of Norlatiadek. Your constellation, Norlatiadek, consists of one hundred local systems and has a headquarters world called Edentia. Norlatiadek is number seventy in the universe of Nebadon. The local universe of Nebadon consists of one hundred constellations and has a capital known as Salvington. The universe of Nebadon is number eighty-four in the minor sector of Ensa.

The minor sector of Ensa consists of one hundred local universes and has a capital called Uminor the third. This minor sector is number three in the major sector of Splandon. Splandon consists of one hundred minor sectors and has a headquarters world called Umajor the fifth. It is the fifth major sector of the superuniverse of Orvonton, the seventh segment of the grand universe. Thus you can locate your planet in the scheme of the organization and administration of the universe of universes.

The grand universe number of your world, Urantia, is 5,342,482,337,666. That is the registry number on Uversa and on Paradise, your number in the catalogue of the inhabited worlds. I know the physical-sphere registry number, but it is of such an extraordinary size that it is of little practical significance to the mortal mind.

Your planet is a member of an enormous cosmos; you belong to a well-nigh infinite family of worlds, but your sphere is just as precisely administered and just as lovingly fostered as if it were the only inhabited world in all existence.

[Presented by a Universal Censor hailing from Uversa.]

SIX

PERSONALITIES OF THE GRAND UNIVERSE

(PAPER 30)

THE personalities and other-than-personal entities now functioning on Paradise and in the grand universe constitute a well-nigh limitless number of living beings. Even the number of major orders and types would stagger the human imagination, let alone the countless subtypes and variations. It is, however, desirable to present something of two basic classifications of living beings—a suggestion of the Paradise classification and an abbreviation of the Uversa Personality Register.

It is not possible to formulate comprehensive and entirely consistent classifications of the personalities of the grand universe because all of the groups are not revealed. It would require numerous additional papers to cover the further revelation required to systematically classify all groups. Such conceptual expansion would hardly be desirable as it would deprive the thinking mortals of the next thousand years of that stimulus to creative speculation which these partially revealed concepts supply. It is best that man not have an overrevelation; it stifles imagination.

1. THE PARADISE CLASSIFICATION OF LIVING BEINGS

Living beings are classified on Paradise in accordance with inherent and attained relationship to the Paradise Deities. During the grand gatherings of the central and superuniverses those present are often grouped in accordance with origin: those of triune origin, or of Trinity attainment; those of dual origin; and those of single origin. It is difficult to interpret the Paradise classification of living beings to the mortal mind, but we are authorized to present the following:

I. *TRIUNE-ORIGIN BEINGS.* Beings created by all three Paradise Deities, either as such or as the Trinity, together with the Trinitized Corps, which designation refers to all groups of trinitized beings, revealed and unrevealed.

A. *The Supreme Spirits.*
1. The Seven Master Spirits.
2. The Seven Supreme Executives.
3. The Seven Orders of Reflective Spirits.

B. *The Stationary Sons of the Trinity.*
1. Trinitized Secrets of Supremacy.
2. Eternals of Days.
3. Ancients of Days.
4. Perfections of Days.
5. Recents of Days.
6. Unions of Days.
7. Faithfuls of Days.
8. Perfectors of Wisdom.
9. Divine Counselors.
10. Universal Censors.

C. *Trinity-origin and Trinitized Beings.*
1. Trinity Teacher Sons.
2. Inspired Trinity Spirits.
3. Havona Natives.
4. Paradise Citizens.
5. Unrevealed Trinity-origin Beings.
6. Unrevealed Deity-trinitized Beings.
7. Trinitized Sons of Attainment.
8. Trinitized Sons of Selection.

    9. Trinitized Sons of Perfection.

   10. Creature-trinitized Sons.

II. *DUAL-ORIGIN BEINGS.* Those of origin in any two of the Paradise Deities or otherwise created by any two beings of direct or indirect descent from the Paradise Deities.

  A. *The Descending Orders.*
1. Creator Sons.
2. Magisterial Sons.
3. Bright and Morning Stars.
4. Father Melchizedeks.
5. The Melchizedeks.
6. The Vorondadeks.
7. The Lanonandeks.
8. Brilliant Evening Stars.
9. The Archangels.
10. Life Carriers.
11. Unrevealed Universe Aids.
12. Unrevealed Sons of God.

  B. *The Stationary Orders.*
1. Abandonters.
2. Susatia.
3. Univitatia.
4. Spironga.
5. Unrevealed Dual-origin Beings.

  C. *The Ascending Orders.*
1. Adjuster-fused Mortals.
2. Son-fused Mortals.
3. Spirit-fused Mortals.
4. Translated Midwayers.
5. Unrevealed Ascenders.

III. *SINGLE-ORIGIN BEINGS.* Those of origin in any one of the Paradise Deities or otherwise created by any one being of direct or indirect descent from the Paradise Deities.

  A. *The Supreme Spirits.*
1. Gravity Messengers.
2. The Seven Spirits of the Havona Circuits.

3. The Twelvefold Adjutants of the Havona Circuits.
4. The Reflective Image Aids.
5. Universe Mother Spirits.
6. The Sevenfold Adjutant Mind-Spirits.
7. Unrevealed Deity-origin Beings.

B. *The Ascending Orders.*
1. Personalized Adjusters.
2. Ascending Material Sons.
3. Evolutionary Seraphim.
4. Evolutionary Cherubim.
5. Unrevealed Ascenders.

C. *The Family of the Infinite Spirit.*
1. Solitary Messengers.
2. Universe Circuit Supervisors.
3. Census Directors.
4. Personal Aids of the Infinite Spirit.
5. Associate Inspectors.
6. Assigned Sentinels.
7. Graduate Guides.
8. Havona Servitals.
9. Universal Conciliators.
10. Morontia Companions.
11. Supernaphim.
12. Seconaphim.
13. Tertiaphim.
14. Omniaphim.
15. Seraphim.
16. Cherubim and Sanobim.
17. Unrevealed Spirit-origin Beings.
18. The Seven Supreme Power Directors.
19. The Supreme Power Centers.
20. The Master Physical Controllers.
21. The Morontia Power Supervisors.

IV. *EVENTUATED TRANSCENDENTAL BEINGS.* There is to be found on Paradise a vast host of transcendental beings whose origin is not ordinarily disclosed to the universes of time and space until they are settled in light and life. These Transcendentalers are neither creators nor creatures; they are the *eventuated* children of divinity, ultimacy, and eternity. These "eventuators" are neither finite nor infinite—they are *absonite;* and absonity is neither infinity nor absoluteness.

These uncreated noncreators are ever loyal to the Paradise Trinity and obedient to the Ultimate. They are existent on four ultimate levels of personality activity and are functional on the seven levels of the absonite in twelve grand divisions consisting of one thousand major working groups of seven classes each. These eventuated beings include the following orders:

1. The Architects of the Master Universe.

2. Transcendental Recorders.

3. Other Transcendentalers.

4. Primary Eventuated Master Force Organizers.

5. Associate Transcendental Master Force Organizers.

God, as a superperson, eventuates; God, as a person, creates; God, as a preperson, fragments; and such an Adjuster fragment of himself evolves the spirit soul upon the material and mortal mind in accordance with the freewill choosing of the personality which has been bestowed upon such a mortal creature by the parental act of God as a Father.

V. *FRAGMENTED ENTITIES OF DEITY.* This order of living existence, originating in the Universal Father, is best typified by the Thought Adjusters, though these entities are by no means the only fragmentations of the prepersonal reality of the First Source and Center. The functions of the other-than-Adjuster fragments are manifold and little known. Fusion with an Adjuster or other such fragment constitutes the creature a *Father-fused being.*

The fragmentations of the premind spirit of the Third Source and Center, though hardly comparable to the Father fragments, should be here recorded. Such entities differ very greatly from Adjusters; they do not as such dwell on Spiritington, nor do they as such traverse the mind-gravity circuits; neither do they indwell mortal creatures during the life in the flesh. They are not prepersonal in the sense that the Adjusters are, but such fragments of premind spirit are bestowed upon certain of the surviving mortals, and fusion therewith constitutes them *Spirit-fused mortals* in contradistinction to Adjuster-fused mortals.

Still more difficult of description is the individualized spirit of a Creator Son, union with which constitutes the creature a *Son-fused mortal*. And there are still other fragmentations of Deity.

VI. *SUPERPERSONAL BEINGS*. There is a vast host of other-than-personal beings of divine origin and of manifold service in the universe of universes. Certain of these beings are resident on the Paradise worlds of the Son; others, like the superpersonal representatives of the Eternal Son, are encountered elsewhere. They are for the most part unmentioned in these narratives, and it would be quite futile to attempt their description to *personal* creatures.

VII. *UNCLASSIFIED AND UNREVEALED ORDERS*. During the present universe age it would not be possible to place all beings, personal or otherwise, within classifications pertaining to the present universe age; nor have all such categories been revealed in these narratives; hence numerous orders have been omitted from these lists. Consider the following:

The Consummator of Universe Destiny.

The Qualified Vicegerents of the Ultimate.

The Unqualified Supervisors of the Supreme.

The Unrevealed Creative Agencies of the Ancients of Days.

Majeston of Paradise.

The Unnamed Reflectivator Liaisons of Majeston.

The Midsonite Orders of the Local Universes.

No especial significance need attach to the listing of these orders together except that none of them appear in the Paradise classification as revealed herein. These are the unclassified few; you have yet to learn of the unrevealed many.

There are spirits: spirit entities, spirit presences, personal spirits, prepersonal spirits, superpersonal spirits, spirit existences, spirit personalities—but neither mortal language nor mortal intellect are adequate. We may however state that there are no personalities of "pure mind"; no entity has personality unless he is endowed with it by God who is spirit. Any mind entity that is not associated with either spiritual or physical energy is not a personality. But in the same sense that there are spirit personalities who have mind there are mind personalities who have spirit. Majeston and his associates are fairly good illustrations of mind-dominated beings, but there are better illustrations of this type of personality unknown to you. There are even whole unrevealed orders of such *mind personalities,* but they are always spirit associated. Certain other unrevealed creatures are what might be termed *mindal- and physical-energy personalities.* This type of being is nonresponsive to spirit gravity but is nonetheless a true personality—is within the Father's circuit.

These papers do not—cannot—even begin to exhaust the story of the living creatures, creators, eventuators, and still-otherwise-existent beings who live and worship and serve in the swarming universes of time and in the central universe of eternity. You mortals are persons; hence we can describe beings who are *personalized,* but how could an *absonitized* being ever be explained to you?

2. The Uversa Personality Register

The divine family of living beings is registered on Uversa in seven grand divisions:

1. The Paradise Deities.
2. The Supreme Spirits.
3. The Trinity-origin Beings.
4. The Sons of God.
5. Personalities of the Infinite Spirit.
6. The Universe Power Directors.
7. The Corps of Permanent Citizenship.

These groups of will creatures are divided into numerous classes and minor subdivisions. The presentation of this classification of the personalities of the grand universe is however chiefly concerned in setting forth those orders of intelligent beings who have been revealed in these narratives, most of whom will be encountered in the ascendant experience of the mortals of time on their progressive climb to Paradise. The following listings make no mention of vast orders of universe beings who carry forward their work apart from the mortal ascension scheme.

I. *THE PARADISE DEITIES.*
1. The Universal Father.
2. The Eternal Son.
3. The Infinite Spirit.

II. *THE SUPREME SPIRITS.*
  1. The Seven Master Spirits.
  2. The Seven Supreme Executives.
  3. The Seven Groups of Reflective Spirits.
  4. The Reflective Image Aids.
  5. The Seven Spirits of the Circuits.
  6. Local Universe Creative Spirits.
  7. Adjutant Mind-Spirits.

III. *THE TRINITY-ORIGIN BEINGS.*
  1. Trinitized Secrets of Supremacy.
  2. Eternals of Days.
  3. Ancients of Days.
  4. Perfections of Days.
  5. Recents of Days.
  6. Unions of Days.
  7. Faithfuls of Days.
  8. Trinity Teacher Sons.
  9. Perfectors of Wisdom.
  10. Divine Counselors.
  11. Universal Censors.
  12. Inspired Trinity Spirits.
  13. Havona Natives.
  14. Paradise Citizens.

IV. *THE SONS OF GOD.*

  A. *Descending Sons.*
  1. Creator Sons—Michaels.
  2. Magisterial Sons—Avonals.
  3. Trinity Teacher Sons—Daynals.
  4. Melchizedek Sons.
  5. Vorondadek Sons.
  6. Lanonandek Sons.
  7. Life Carrier Sons.

  B. *Ascending Sons.*
  1. Father-fused Mortals.
  2. Son-fused Mortals.
  3. Spirit-fused Mortals.
  4. Evolutionary Seraphim.

     5. Ascending Material Sons.

     6. Translated Midwayers.

     7. Personalized Adjusters.

C. *Trinitized Sons.*

1. Mighty Messengers.
2. Those High in Authority.
3. Those without Name and Number.
4. Trinitized Custodians.
5. Trinitized Ambassadors.
6. Celestial Guardians.
7. High Son Assistants.
8. Ascender-trinitized Sons.
9. Paradise-Havona-trinitized Sons.
10. Trinitized Sons of Destiny.

V. *PERSONALITIES OF THE INFINITE SPIRIT.*

A. *Higher Personalities of the Infinite Spirit.*

1. Solitary Messengers.
2. Universe Circuit Supervisors.
3. Census Directors.
4. Personal Aids of the Infinite Spirit.
5. Associate Inspectors.
6. Assigned Sentinels.
7. Graduate Guides.

B. *The Messenger Hosts of Space.*

1. Havona Servitals.
2. Universal Conciliators.
3. Technical Advisers.
4. Custodians of Records on Paradise.
5. Celestial Recorders.
6. Morontia Companions.
7. Paradise Companions.

C. *The Ministering Spirits.*

1. Supernaphim.
2. Seconaphim.
3. Tertiaphim.
4. Omniaphim.
5. Seraphim.

6. Cherubim and Sanobim.
7. Midwayers.

VI. *THE UNIVERSE POWER DIRECTORS.*
  A. *The Seven Supreme Power Directors.*

  B. *Supreme Power Centers.*
    1. Supreme Center Supervisors.
    2. Havona Centers.
    3. Superuniverse Centers.
    4. Local Universe Centers.
    5. Constellation Centers.
    6. System Centers.
    7. Unclassified Centers.

  C. *Master Physical Controllers.*
    1. Associate Power Directors.
    2. Mechanical Controllers.
    3. Energy Transformers.
    4. Energy Transmitters.
    5. Primary Associators.
    6. Secondary Dissociators.
    7. Frandalanks and Chronoldeks.

  D. *Morontia Power Supervisors.*
    1. Circuit Regulators.
    2. System Co-ordinators.
    3. Planetary Custodians.
    4. Combined Controllers.
    5. Liaison Stabilizers.
    6. Selective Assorters.
    7. Associate Registrars.

VII. *THE CORPS OF PERMANENT CITIZENSHIP.*
    1. The Planetary Midwayers.
    2. The Adamic Sons of the Systems.
    3. The Constellation Univitatia.
    4. The Local Universe Susatia.
    5. Spirit-fused Mortals of the Local Universes.
    6. The Superuniverse Abandonters.
    7. Son-fused Mortals of the Superuniverses.

8. The Havona Natives.
9. Natives of the Paradise Spheres of the Spirit.
10. Natives of the Father's Paradise Spheres.
11. The Created Citizens of Paradise.
12. Adjuster-fused Mortal Citizens of Paradise.

This is the working classification of the personalities of the universes as they are of record on the headquarters world of Uversa.

COMPOSITE PERSONALITY GROUPS. There are on Uversa the records of numerous additional groups of intelligent beings, beings that are also closely related to the organization and administration of the grand universe. Among such orders are the following three composite personality groups:

A. *The Paradise Corps of the Finality.*
1. The Corps of Mortal Finaliters.
2. The Corps of Paradise Finaliters.
3. The Corps of Trinitized Finaliters.
4. The Corps of Conjoint Trinitized Finaliters.
5. The Corps of Havona Finaliters.
6. The Corps of Transcendental Finaliters.
7. The Corps of Unrevealed Sons of Destiny.

The Mortal Corps of the Finality is dealt with in the next and final paper of this series.

B. *The Universe Aids.*
1. Bright and Morning Stars.
2. Brilliant Evening Stars.
3. Archangels.
4. Most High Assistants.
5. High Commissioners.
6. Celestial Overseers.
7. Mansion World Teachers.

On all headquarters worlds of both local and superuniverses, provision is made for these beings who are engaged in specific missions for the Creator Sons, the local universe rulers. We welcome these *Universe Aids* on Uversa, but we have no jurisdiction over them. Such emissaries prosecute their work and carry on their observations under authority of the Creator Sons. Their activities are more fully described in the narrative of your local universe.

C. *The Seven Courtesy Colonies.*
1. Star Students.
2. Celestial Artisans.
3. Reversion Directors.
4. Extension-school Instructors.
5. The Various Reserve Corps.
6. Student Visitors.
7. Ascending Pilgrims.

These seven groups of beings will be found thus organized and governed on all headquarters worlds from the local systems up to the capitals of the superuniverses, particularly the latter. The capitals of the seven superuniverses are the meeting places for almost all classes and orders of intelligent beings. With the exception of numerous groups of Paradise-Havoners, here the will creatures of every phase of existence may be observed and studied.

3. THE COURTESY COLONIES

The seven courtesy colonies sojourn on the architectural spheres for a longer or shorter time while engaged in the furtherance of their missions and in the execution of their special assignments. Their work may be described as follows:

1. *The Star Students,* the celestial astronomers, choose to work on spheres like Uversa because such specially constructed worlds are unusually favorable for their observations and calculations. Uversa is favorably situated for the work of this colony, not only because of its central location, but also because there are no gigantic living or dead suns near at hand to disturb the energy currents. These students are not in any manner organically connected with the affairs of the superuniverse; they are merely guests.

The astronomical colony of Uversa contains individuals from many near-by realms, from the central universe, and even from Norlatiadek. Any being on any world in any system of any universe may become a star student, may aspire to join some corps of celestial astronomers. The only requisites are: continuing life and sufficient knowledge of the worlds of space, especially their physical laws of evolution and control. Star students are not required to serve eternally in this corps, but no one admitted to this group may withdraw under one millennium of Uversa time.

The star-observer colony of Uversa now numbers over one million. These astronomers come and go, though some remain for comparatively long periods. They carry on their work with the aid of a multitude of mechanical instruments and physical appliances; they are also greatly assisted by the Solitary Messengers and other spirit explorers. These celestial astronomers make constant use of the living energy transformers and transmitters, as well as of the reflective personalities, in their work of star study and space survey. They study all forms and phases of space material and energy manifestations, and they are just as much interested in force function as in stellar phenomena; nothing in all space escapes their scrutiny.

Similar astronomer colonies are to be found on the sector headquarters worlds of the superuniverse as well as on the architectural capitals of the local universes and their administrative subdivisions. Except on Paradise, knowledge is not inherent; understanding of the physical universe is largely dependent on observation and research.

2. *The Celestial Artisans* serve throughout the seven superuniverses. Ascending mortals have their initial contact with these groups in the morontia career of the local universe in connection with which these artisans will be more fully discussed.

3. *The Reversion Directors* are the promoters of relaxation and humor—reversion to past memories. They are of great service in the practical operation of the ascending scheme of mortal progression, especially during the earlier phases of morontia transition and spirit experience. Their story belongs to the narrative of the mortal career in the local universe.

4. *Extension-School Instructors.* The next higher residential world of the ascendant career always maintains a strong corps of teachers on the world just below, a sort of preparatory school for the progressing residents of that sphere; this is a phase of the ascendant scheme for advancing the pilgrims of time. These schools, their methods of instruction and examinations, are wholly unlike anything which you essay to conduct on Urantia.

The entire ascendant plan of mortal progression is characterized by the practice of giving out to other beings new truth and experience just as soon as acquired. You work your way through the long school of Paradise attainment by serving as teachers to those pupils just behind you in the scale of progression.

5. *The Various Reserve Corps.* Vast reserves of beings not under our immediate supervision are mobilized on Uversa as the reserve-corps colony. There are seventy primary divisions of this colony on Uversa, and it is a liberal education to be permitted to spend a season with these extraordinary personalities. Similar general reserves are maintained on Salvington and other universe capitals; they are dispatched on active service on the requisition of their respective group directors.

6. *The Student Visitors.* From all the universe a constant stream of celestial visitors pours through the various headquarters worlds. As individuals and as classes these various types of beings flock in upon us as observers, exchange pupils, and student helpers. On Uversa, at present, there are over one billion persons in this courtesy colony. Some of these visitors may tarry a day, others may remain a year, all dependent on the nature of their mission. This colony contains almost every class of universe beings except Creator personalities and morontia mortals.

Morontia mortals are student visitors only within the confines of the local universe of their origin. They may visit in a superuniverse capacity only after they have attained spirit status. Fully one half of our visitor colony consists of "stopovers," beings en route elsewhere who pause to visit the Orvonton capital. These personalities may be executing a universe assignment, or they may be enjoying a period of leisure—freedom from assignment. The privilege of intrauniverse travel and observation is a part of the career of all ascending beings. The human desire to travel and observe new peoples and worlds will be fully gratified during the long and eventful climb to Paradise through the local, super-, and central universes.

7. *The Ascending Pilgrims.* As the ascending pilgrims are assigned to various services in connection with their Paradise progression, they are domiciled as a courtesy colony on the various headquarters spheres. While functioning here and there throughout a superuniverse, such groups are largely self-governing. They are an ever-shifting colony embracing all orders of evolutionary mortals and their ascending associates.

4. THE ASCENDING MORTALS

While the mortal survivors of time and space are denominated *ascending pilgrims* when accredited for the progressive ascent to Paradise, these evolutionary creatures occupy such an important place in these narratives that we here desire to present a synopsis of the following seven stages of the ascending universe career:

1. Planetary Mortals.

2. Sleeping Survivors.

3. Mansion World Students.

4. Morontia Progressors.

5. Superuniverse Wards.

6. Havona Pilgrims.

7. Paradise Arrivals.

The following narrative presents the universe career of an Adjuster-indwelt mortal. The Son- and Spirit-fused mortals share portions of this career, but we have elected to tell this story as it pertains to the Adjuster-fused mortals, for such a destiny may be anticipated by all of the human races of Urantia.

1. *Planetary Mortals.* Mortals are all animal-origin evolutionary beings of ascendant potential. In origin, nature, and destiny these various groups and types of human beings are not wholly unlike the Urantia peoples. The human races of each world receive the same ministry of the Sons of God and enjoy the presence of the ministering spirits of time. After natural death all types of ascenders fraternize as one morontia family on the mansion worlds.

2. *Sleeping Survivors.* All mortals of survival status, in the custody of personal guardians of destiny, pass through the portals of natural death and, on the third period, personalize on the mansion worlds. Those accredited beings who have, for any reason, been unable to attain that level of intelligence mastery and endowment of spirituality which would entitle them to personal guardians, cannot thus immediately and directly go to the mansion worlds. Such surviving souls must rest in unconscious sleep until the judgment day of a new epoch, a new dispensation, the coming of a Son of God to call the rolls of the age and adjudicate the realm, and this is the general practice throughout all Nebadon. It was said of Christ Michael that, when he ascended on high at the conclusion of his work on earth, "He led a great multitude of captives." And these captives were the sleeping survivors from the days of Adam to the day of the Master's resurrection on Urantia.

The passing of time is of no moment to sleeping mortals; they are wholly unconscious and oblivious to the length of their rest. On reassembly of personality at the end of an age, those who have slept five thousand years will react no differently than those who have rested five days. Aside from this time delay these survivors pass on through the ascension regime identically with those who avoid the longer or shorter sleep of death.

These dispensational classes of world pilgrims are utilized for group morontia activities in the work of the local universes. There is a great advantage in the mobilization of such enormous groups; they are thus kept together for long periods of effective service.

3. *Mansion World Students.* All surviving mortals who re-awaken on the mansion worlds belong to this class.

The physical body of mortal flesh is not a part of the reassembly of the sleeping survivor; the physical body has returned to dust. The seraphim of assignment sponsors the new body, the morontia form, as the new life vehicle for the immortal soul and for the indwelling of the returned Adjuster. The Adjuster is the custodian of the spirit transcript of the mind of the sleeping survivor. The assigned seraphim is the keeper of the surviving identity—the immortal soul—as far as it has evolved. And when these two, the Adjuster and the seraphim, reunite their personality trusts, the new individual constitutes the resurrection of the old personality, the survival of the evolving morontia identity of the soul. Such a reassociation of soul and Adjuster is quite properly called a resurrection, a reassembly of personality factors; but even this does not entirely explain the reappearance of the surviving *personality*. Though you will probably never understand the fact of such an inexplicable transaction, you will sometime experientially know the truth of it if you do not reject the plan of mortal survival.

The plan of initial mortal detention on seven worlds of progressive training is nearly universal in Orvonton. In each local system of approximately one thousand inhabited planets there are seven mansion worlds, usually satellites or subsatellites of the system capital. They are the receiving worlds for the majority of ascending mortals.

Sometimes all training worlds of mortal residence are called universe "mansions," and it was to such spheres that Jesus alluded when he said: "In my Father's house are many mansions." From here on, within a given group of spheres like the mansion worlds, ascenders will progress individually from one sphere to another and from one phase of life to another, but they will always advance from one stage of universe study to another in class formation.

4. *Morontia Progressors.* From the mansion worlds on up through the spheres of the system, constellation, and the universe, mortals are classed as morontia progressors; they are traversing the transition spheres of mortal ascension. As the ascending mortals progress from the lower to the higher of the morontia worlds, they serve on countless assignments in association with their teachers and in company with their more advanced and senior brethren.

Morontia progression pertains to continuing advancement of intellect, spirit, and personality form. Survivors are still three-natured beings. Throughout the entire morontia experience they are wards of the local universe. The regime of the superuniverse does not function until the spirit career begins.

Mortals acquire real spirit identity just before they leave the local universe headquarters for the receiving worlds of the minor sectors of the superuniverse. Passing from the final morontia stage to the first or lowest spirit status is but a slight transition. The mind, personality, and character are unchanged by such an advance; only does the form undergo modification. But the spirit form is just as real as the morontia body, and it is equally discernible.

Before departing from their native local universes for the superuniverse receiving worlds, the mortals of time are recipients of spirit confirmation from the Creator Son and the local universe Mother Spirit. From this point on, the status of the ascending mortal is forever settled. Superuniverse wards have never been known to go astray. Ascending seraphim are also advanced in angelic standing at the time of their departure from the local universes.

5. *Superuniverse Wards.* All ascenders arriving on the training worlds of the superuniverses become the wards of the Ancients of Days; they have traversed the morontia life of the local universe and are now accredited spirits. As young spirits they begin the ascension of the superuniverse system of training and culture, extending from the receiving spheres of their minor sector in through the study worlds of the ten major sectors and on to the higher cultural spheres of the superuniverse headquarters.

There are three orders of student spirits in accordance with their sojourn upon the minor sector, major sectors, and the superuniverse headquarters worlds of spirit progression. As morontia ascenders studied and worked on the worlds of the local universe, so spirit ascenders continue to master new worlds while they practice at giving out to others that which they have imbibed at the experiential founts of wisdom. But going to school as a spirit being in the superuniverse career is very unlike anything that has ever entered the imaginative realms of the material mind of man.

Before leaving the superuniverse for Havona, these ascending spirits receive the same thorough course in superuniverse management that they received during their morontia experience in local universe supervision. Before spirit mortals reach Havona, their chief study, but not exclusive occupation, is the mastery of local and superuniverse administration. The reason for all of this experience is not now fully apparent, but no doubt such training is wise and necessary in view of their possible future destiny as members of the Corps of the Finality.

The superuniverse regime is not the same for all ascending mortals. They receive the same general education, but special groups and classes are carried through special courses of instruction and are put through specific courses of training.

6. *Havona Pilgrims*. When spirit development is complete, even though not replete, then the surviving mortal prepares for the long flight to Havona, the haven of evolutionary spirits. On earth you were a creature of flesh and blood; through the local universe you were a morontia being; through the superuniverse you were an evolving spirit; with your arrival on the receiving worlds of Havona your spiritual education begins in reality and in earnest; your eventual appearance on Paradise will be as a perfected spirit.

The journey from the superuniverse headquarters to the Havona receiving spheres is always made alone. From now on no more class or group instruction will be administered. You are through with the technical and administrative training of the evolutionary worlds of time and space. Now begins your *personal education*, your individual spiritual training. From first to last, throughout all Havona, the instruction is personal and threefold in nature: intellectual, spiritual, and experiential.

The first act of your Havona career will be to recognize and thank your transport seconaphim for the long and safe journey. Then you are presented to those beings who will sponsor your early Havona activities. Next you go to register your arrival and prepare your message of thanksgiving and adoration for dispatch to the Creator Son of your local universe, the universe Father who made possible your sonship career. This concludes the formalities of the Havona arrival; whereupon you are accorded a long period of leisure for free observation, and this affords opportunity for looking up your friends, fellows, and associates of the long ascension experience. You may also consult the broadcasts to ascertain who of your fellow pilgrims have departed for Havona since the time of your leaving Uversa.

The fact of your arrival on the receiving worlds of Havona will be duly transmitted to the headquarters of your local universe and personally conveyed to your seraphic guardian, wherever that seraphim may chance to be.

The ascendant mortals have been thoroughly trained in the affairs of the evolutionary worlds of space; now they begin their long and profitable contact with the created spheres of perfection. What a preparation for some future work is afforded by this combined, unique, and extraordinary experience! But I cannot tell you about Havona; you must see these worlds to appreciate their glory or to understand their grandeur.

7. *Paradise Arrivals.* On reaching Paradise with residential status, you begin the progressive course in divinity and absonity. Your residence on Paradise signifies that you have found God, and that you are to be mustered into the Mortal Corps of the Finality. Of all the creatures of the grand universe, only those who are Father fused are mustered into the Mortal Corps of the Finality. Only such individuals take the finaliter oath. Other beings of Paradise perfection or attainment may be temporarily attached to this finality corps, but they are not of eternal assignment to the unknown and unrevealed mission of this accumulating host of the evolutionary and perfected veterans of time and space.

Paradise arrivals are accorded a period of freedom, after which they begin their associations with the seven groups of the primary supernaphim. They are designated Paradise graduates when they have finished their course with the conductors of worship and then, as finaliters, are assigned on observational and co-operative service to the ends of the far-flung creation. As yet there seems to be no specific or settled employment for the Mortal Corps of Finaliters, though they serve in many capacities on worlds settled in light and life.

If there should be no future or unrevealed destiny for the Mortal Corps of the Finality, the present assignment of these ascendant beings would be altogether adequate and glorious. Their present destiny wholly justifies the universal plan of evolutionary ascent. But the future ages of the evolution of the spheres of outer space will undoubtedly further elaborate, and with more repleteness divinely illuminate, the wisdom and loving-kindness of the Gods in the execution of their divine plan of human survival and mortal ascension.

This narrative, together with what has been revealed to you and with what you may acquire in connection with instruction respecting your own world, presents an outline of the career of an ascending mortal. The story varies considerably in the different superuniverses, but this recital affords a glimpse of the average plan of mortal progression as it is operative in the local universe of Nebadon and in the seventh segment of the grand universe, the superuniverse of Orvonton.

[Sponsored by a Mighty Messenger from Uversa.]

SEVEN

THE EVOLUTION
OF LOCAL UNIVERSES

(PAPER 32)

A LOCAL universe is the handiwork of a Creator Son of the Paradise order of Michael. It comprises one hundred constellations, each embracing one hundred systems of inhabited worlds. Each system will eventually contain approximately one thousand inhabited spheres.

These universes of time and space are all evolutionary. The creative plan of the Paradise Michaels always proceeds along the path of gradual evolvement and progressive development of the physical, intellectual, and spiritual natures and capacities of the manifold creatures who inhabit the varied orders of spheres comprising such a local universe.

Urantia belongs to a local universe whose sovereign is the God-man of Nebadon, Jesus of Nazareth and Michael of Salvington. And all of Michael's plans for this local universe were fully approved by the Paradise Trinity before he ever embarked upon the supreme adventure of space.

The Sons of God may choose the realms of their creator activities, but these material creations were originally projected and planned by the Paradise Architects of the Master Universe.

1. Physical Emergence of Local Universes

The preuniverse manipulations of space-force and the primordial energies are the work of the Paradise Master Force Organizers; but in the superuniverse domains, when emergent energy becomes responsive to local or linear gravity, they retire in favor of the power directors of the superuniverse concerned.

These power directors function alone in the prematerial and postforce phases of a local universe creation. There is no opportunity for a Creator Son to begin universe organization until the power directors have effected the mobilization of the space-energies sufficiently to provide a material foundation—literal suns and material spheres—for the emerging universe.

The local universes are all approximately of the same energy potential, though they differ greatly in physical dimensions and may vary in visible-matter content from time to time. The power charge and potential-matter endowment of a local universe are determined by the manipulations of the power directors and their predecessors as well as by the Creator Son's activities and by the endowment of the inherent physical control possessed by his creative associate.

The energy charge of a local universe is approximately one one-hundred-thousandth of the force endowment of its superuniverse. In the case of Nebadon, your local universe, the mass materialization is a trifle less. Physically speaking, Nebadon possesses all of the physical endowment of energy and matter that may be found in any of the Orvonton local creations. The only physical limitation upon the developmental expansion of the Nebadon universe consists in the quantitative charge of space-energy held captive by the gravity control of the associated powers and personalities of the combined universe mechanism.

When energy-matter has attained a certain stage in mass materialization, a Paradise Creator Son appears upon the scene, accompanied by a Creative Daughter of the Infinite Spirit. Simultaneously with the arrival of the Creator Son, work is begun upon the architectural sphere which is to become the headquarters world of the projected local universe. For long ages such a local creation evolves, suns become stabilized, planets form and swing into their orbits, while the work of creating the architectural worlds which are to serve as constellation headquarters and system capitals continues.

2. UNIVERSE ORGANIZATION

Creator Sons are preceded in universe organization by the power directors and other beings originating in the Third Source and Center. From the energies of space, thus previously organized, Michael, your Creator Son, established the inhabited realms of the universe of Nebadon and ever since has been painstakingly devoted to their administration. From pre-existent energy these divine Sons materialize visible matter, project living creatures, and with the co-operation of the universe presence of the Infinite Spirit, create a diverse retinue of spirit personalities.

These power directors and energy controllers who long preceded the Creator Son in the preliminary physical work of universe organization later serve in magnificent liaison with this Universe Son, forever remaining in associated control of those energies which they originally organized and circuitized. On Salvington there now function the same one hundred power centers who co-operated with your Creator Son in the original formation of this local universe.

The first completed act of physical creation in Nebadon consisted in the organization of the headquarters world, the architectural sphere of Salvington, with its satellites. From the time of the initial moves of the power centers and physical controllers to the arrival of the living staff on the completed spheres of Salvington, there intervened a little over one billion years of your present planetary time. The construction of Salvington was immediately followed by the creation of the one hundred headquarters worlds of the projected constellations and the ten thousand headquarters spheres of the projected local systems of planetary control and administration, together with their architectural satellites. Such architectural worlds are designed to accommodate both physical and spiritual personalities as well as the intervening morontia or transition stages of being.

Salvington, the headquarters of Nebadon, is situated at the exact energy-mass center of the local universe. But your local universe is not a single astronomic system, though a large system does exist at its physical center.

Salvington is the personal headquarters of Michael of Nebadon, but he will not always be found there. While the smooth functioning of your local universe no longer requires the fixed presence of the Creator Son at the capital sphere, this was not true of the earlier epochs of physical organization. A Creator Son is unable to leave his headquarters world until such a time as gravity stabilization of the realm has been effected through the materialization of sufficient energy to enable the various circuits and systems to counterbalance one another by mutual material attraction.

Presently, the physical plan of a universe is completed, and the Creator Son, in association with the Creative Spirit, projects his plan of life creation; whereupon does this representation of the Infinite Spirit begin her universe function as a distinct creative personality. When this first creative act is formulated and executed, there springs into being the Bright and Morning Star, the personification of this initial creative concept of identity and ideal of divinity. This is the chief executive of the universe, the personal associate of the Creator Son, one like him in all aspects of character, though markedly limited in the attributes of divinity.

And now that the right-hand helper and chief executive of the Creator Son has been provided, there ensues the bringing into existence of a vast and wonderful array of diverse creatures. The sons and daughters of the local universe are forthcoming, and soon thereafter the government of such a creation is provided, extending from the supreme councils of the universe to the fathers of the constellations and the sovereigns of the local systems—the aggregations of those worlds which are designed subsequently to become the homes of the varied mortal races of will creatures; and each of these worlds will be presided over by a Planetary Prince.

And then, when such a universe has been so completely organized and so repletely manned, does the Creator Son enter into the Father's proposal to create mortal man in their divine image.

The organization of planetary abodes is still progressing in Nebadon, for this universe is, indeed, a young cluster in the starry and planetary realms of Orvonton. At the last registry there were 3,840,101 inhabited planets in Nebadon, and Satania, the local system of your world, is fairly typical of other systems.

Satania is not a uniform physical system, a single astronomic unit or organization. Its 619 inhabited worlds are located in over five hundred different physical systems. Only five have more than two inhabited worlds, and of these only one has four peopled planets, while there are forty-six having two inhabited worlds.

The Satania system of inhabited worlds is far removed from Uversa and that great sun cluster which functions as the physical or astronomic center of the seventh superuniverse. From Jerusem, the headquarters of Satania, it is over two hundred thousand light-years to the physical center of the superuniverse of Orvonton, far, far away in the dense diameter of the Milky Way. Satania is on the periphery of the local universe, and Nebadon is now well out towards the edge of Orvonton. From the outermost system of inhabited worlds to the center of the superuniverse is a trifle less than two hundred and fifty thousand light-years.

The universe of Nebadon now swings far to the south and east in the superuniverse circuit of Orvonton. The nearest neighboring universes are: Avalon, Henselon, Sanselon, Portalon, Wolvering, Fanoving, and Alvoring.

But the evolution of a local universe is a long narrative. Papers dealing with the superuniverse introduce this subject, those of this section, treating of the local creations, continue it, while those to follow, touching upon the history and destiny of Urantia, complete the story. But you can adequately comprehend the destiny of the mortals of such a local creation only by a perusal of the narratives of the life and teachings of your Creator Son as he once lived the life of man, in the likeness of mortal flesh, on your own evolutionary world.

3. THE EVOLUTIONARY IDEA

The only creation that is perfectly settled is Havona, the central universe, which was made directly by the thought of the Universal Father and the word of the Eternal Son. Havona is an existential, perfect, and replete universe, surrounding the home of the eternal Deities, the center of all things. The creations of the seven superuniverses are finite, evolutionary, and consistently progressive.

The physical systems of time and space are all evolutionary in origin. They are not even physically stabilized until they are swung into the settled circuits of their superuniverses. Neither is a local universe settled in light and life until its physical possibilities of expansion and development have been exhausted, and until the spiritual status of all its inhabited worlds has been forever settled and stabilized.

Except in the central universe, perfection is a progressive attainment. In the central creation we have a pattern of perfection, but all other realms must attain that perfection by the methods established for the advancement of those particular worlds or universes. And an almost infinite variety characterizes the plans of the Creator Sons for organizing, evolving, disciplining, and settling their respective local universes.

With the exception of the deity presence of the Father, every local universe is, in a certain sense, a duplication of the administrative organization of the central or pattern creation. Although the Universal Father is personally present in the residential universe, he does not indwell the minds of the beings originating in that universe as he does literally dwell with the souls of the mortals of time and space. There seems to be an all-wise compensation in the adjustment and regulation of the spiritual affairs of the far-flung creation. In the central universe the Father is personally present as such but absent in the minds of the children of that perfect creation; in the universes of space the Father is absent in person, being represented by his Sovereign Sons, while he is intimately present in the minds of his mortal children, being spiritually represented by the prepersonal presence of the Mystery Monitors that reside in the minds of these will creatures.

On the headquarters of a local universe there reside all those creator and creative personalities who represent self-contained authority and administrative autonomy except the personal presence of the Universal Father. In the local universe there are to be found something of everyone and someone of almost every class of intelligent beings existing in the central universe except the Universal Father. Although the Universal Father is not personally present in a local universe, he is personally represented by its Creator Son, sometime vicegerent of God and subsequently supreme and sovereign ruler in his own right.

The farther down the scale of life we go, the more difficult it becomes to locate, with the eye of faith, the invisible Father. The lower creatures—and sometimes even the higher personalities—find it difficult always to envisage the Universal Father in his Creator Sons. And so, pending the time of their spiritual exaltation, when perfection of development will enable them to see God in person, they grow weary in progression, entertain spiritual doubts, stumble into confusion, and thus isolate themselves from the progressive spiritual aims of their time and universe. In this way they lose the ability to see the Father when beholding the Creator Son. The surest safeguard for the creature throughout the long struggle to attain the Father, during this time when inherent conditions make such attainment impossible, is tenaciously to hold on to the truth-fact of the Father's presence in his Sons. Literally and figuratively, spiritually and personally, the Father and the Sons are one. It is a fact: He who has seen a Creator Son has seen the Father.

The personalities of a given universe are settled and dependable, at the start, only in accordance with their degree of kinship to Deity. When creature origin departs sufficiently far from the original and divine Sources, whether we are dealing with the Sons of God or the creatures of ministry belonging to the Infinite Spirit, there is an increase in the possibility of disharmony, confusion, and sometimes rebellion—sin.

Excepting perfect beings of Deity origin, all will creatures in the superuniverses are of evolutionary nature, beginning in lowly estate and climbing ever upward, in reality inward. Even highly spiritual personalities continue to ascend the scale of life by progressive translations from life to life and from sphere to sphere. And in the case of those who entertain the Mystery Monitors, there is indeed no limit to the possible heights of their spiritual ascent and universe attainment.

The perfection of the creatures of time, when finally achieved, is wholly an acquirement, a bona fide personality possession. While the elements of grace are freely admixed, nevertheless, the creature attainments are the result of individual effort and actual living, personality reaction to the existing environment.

The fact of animal evolutionary origin does not attach stigma to any personality in the sight of the universe as that is the exclusive method of producing one of the two basic types of finite intelligent will creatures. When the heights of perfection and eternity are attained, all the more honor to those who began at the bottom and joyfully climbed the ladder of life, round by round, and who, when they do reach the heights of glory, will have gained a personal experience which embodies an actual knowledge of every phase of life from the bottom to the top.

In all this is shown the wisdom of the Creators. It would be just as easy for the Universal Father to make all mortals perfect beings, to impart perfection by his divine word. But that would deprive them of the wonderful experience of the adventure and training associated with the long and gradual inward climb, an experience to be had only by those who are so fortunate as to begin at the very bottom of living existence.

In the universes encircling Havona there are provided only a sufficient number of perfect creatures to meet the need for pattern teacher guides for those who are ascending the evolutionary scale of life. The experiential nature of the evolutionary type of personality is the natural cosmic complement of the ever-perfect natures of the Paradise-Havona creatures. In reality, both perfect and perfected creatures are incomplete as regards finite totality. But in the complemental association of the existentially perfect creatures of the Paradise-Havona system with the experientially perfected finaliters ascending from the evolutionary universes, both types find release from inherent limitations and thus may conjointly attempt to reach the sublime heights of the ultimate of creature status.

These creature transactions are the universe repercussions of actions and reactions within the Sevenfold Deity, wherein the eternal divinity of the Paradise Trinity is conjoined with the evolving divinity of the Supreme Creators of the time-space universes in, by, and through the power-actualizing Deity of the Supreme Being.

The divinely perfect creature and the evolutionary perfected creature are equal in degree of divinity potential, but they differ in kind. Each must depend on the other to attain supremacy of service. The evolutionary superuniverses depend on perfect Havona to provide the final training for their ascending citizens, but so does the perfect central universe require the existence of the perfecting superuniverses to provide for the full development of its descending inhabitants.

The two prime manifestations of finite reality, innate perfection and evolved perfection, be they personalities or universes, are co-ordinate, dependent, and integrated. Each requires the other to achieve completion of function, service, and destiny.

4. God's Relation to a Local Universe

Do not entertain the idea that, since the Universal Father has delegated so much of himself and his power to others, he is a silent or inactive member of the Deity partnership. Aside from personality domains and Adjuster bestowal, he is apparently the least active of the Paradise Deities in that he allows his Deity co-ordinates, his Sons, and numerous created intelligences to perform so much in the carrying out of his eternal purpose. He is the silent member of the creative trio only in that he never does aught which any of his co-ordinate or subordinate associates can do.

God has full understanding of the need of every intelligent creature for function and experience, and therefore, in every situation, be it concerned with the destiny of a universe or the welfare of the humblest of his creatures, God retires from activity in favor of the galaxy of creature and Creator personalities who inherently intervene between himself and any given universe situation or creative event. But notwithstanding this retirement, this exhibition of infinite co-ordination, there is on God's part an actual, literal, and personal participation in these events by and through these ordained agencies and personalities. The Father is working in and through all these channels for the welfare of all his far-flung creation.

As regards the policies, conduct, and administration of a local universe, the Universal Father acts in the person of his Creator Son. In the interrelationships of the Sons of God, in the group associations of the personalities of origin in the Third Source and Center, or in the relationship between any other creatures, such as human beings—as concerns such associations the Universal Father never intervenes. The law of the Creator Son, the rule of the Constellation Fathers, the System Sovereigns, and the Planetary Princes—the ordained policies and procedures for that universe—always prevail. There is no division of authority; never is there a cross working of divine power and purpose. The Deities are in perfect and eternal unanimity.

The Creator Son rules supreme in all matters of ethical associations, the relations of any division of creatures to any other class of creatures or of two or more individuals within any given group; but such a plan does not mean that the Universal Father may not in his own way intervene and do aught that pleases the divine mind with any *individual creature* throughout all creation, as pertains to that individual's present status or future prospects and as concerns the Father's eternal plan and infinite purpose.

In the mortal will creatures the Father is actually present in the indwelling Adjuster, a fragment of his prepersonal spirit; and the Father is also the source of the personality of such a mortal will creature.

These Thought Adjusters, the bestowals of the Universal Father, are comparatively isolated; they indwell human minds but have no discernible connection with the ethical affairs of a local creation. They are not directly co-ordinated with the seraphic service nor with the administration of systems, constellations, or a local universe, not even with the rule of a Creator Son, whose will is the supreme law of his universe.

The indwelling Adjusters are one of God's separate but unified modes of contact with the creatures of his all but infinite creation. Thus does he who is invisible to mortal man manifest his presence, and could he do so, he would show himself to us in still other ways, but such further revelation is not divinely possible.

We can see and understand the mechanism whereby the Sons enjoy intimate and complete knowledge regarding the universes of their jurisdiction; but we cannot fully comprehend the methods whereby God is so fully and personally conversant with the details of the universe of universes, although we at least can recognize the avenue whereby the Universal Father can receive information regarding, and manifest his presence to, the beings of his immense creation. Through the personality circuit the Father is cognizant—has personal knowledge—of all the thoughts and acts of all the beings in all the systems of all the universes of all creation. Though we cannot fully grasp this technique of God's communion with his children, we can be strengthened in the assurance that the "Lord knows his children," and that of each one of us "he takes note where we were born."

In your universe and in your heart the Universal Father is present, spiritually speaking, by one of the Seven Master Spirits of central abode and, specifically, by the divine Adjuster who lives and works and waits in the depths of the mortal mind.

God is not a self-centered personality; the Father freely distributes himself to his creation and to his creatures. He lives and acts, not only in the Deities, but also in his Sons, whom he intrusts with the doing of everything that it is divinely possible for them to do. The Universal Father has truly divested himself of every function which it is possible for another being to perform. And this is just as true of mortal man as of the Creator Son who rules in God's stead at the headquarters of a local universe. Thus we behold the outworking of the ideal and infinite love of the Universal Father.

In this universal bestowal of himself we have abundant proof of both the magnitude and the magnanimity of the Father's divine nature. If God has withheld aught of himself from the universal creation, then of that residue he is in lavish generosity bestowing the Thought Adjusters upon the mortals of the realms, the Mystery Monitors of time, who so patiently indwell the mortal candidates for life everlasting.

The Universal Father has poured out himself, as it were, to make all creation rich in personality possession and potential spiritual attainment. God has given us himself that we may be like him, and he has reserved for himself of power and glory only that which is necessary for the maintenance of those things for the love of which he has thus divested himself of all things else.

5. THE ETERNAL AND DIVINE PURPOSE

There is a great and glorious purpose in the march of the universes through space. All of your mortal struggling is not in vain. We are all part of an immense plan, a gigantic enterprise, and it is the vastness of the undertaking that renders it impossible to see very much of it at any one time and during any one life. We are all a part of an eternal project which the Gods are supervising and outworking. The whole marvelous and universal mechanism moves on majestically through space to the music of the meter of the infinite thought and the eternal purpose of the First Great Source and Center.

The eternal purpose of the eternal God is a high spiritual ideal. The events of time and the struggles of material existence are but the transient scaffolding which bridges over to the other side, to the promised land of spiritual reality and supernal existence. Of course, you mortals find it difficult to grasp the idea of an eternal purpose; you are virtually unable to comprehend the thought of eternity, something never beginning and never ending. Everything familiar to you has an end.

As regards an individual life, the duration of a realm, or the chronology of any connected series of events, it would seem that we are dealing with an isolated stretch of time; everything seems to have a beginning and an end. And it would appear that a series of such experiences, lives, ages, or epochs, when successively arranged, constitutes a straightaway drive, an isolated event of time flashing momentarily across the infinite face of eternity. But when we look at all this from behind the scenes, a more comprehensive view and a more complete understanding suggest that such an explanation is inadequate, disconnected, and wholly unsuited properly to account for, and otherwise to correlate, the transactions of time with the underlying purposes and basic reactions of eternity.

To me it seems more fitting, for purposes of explanation to the mortal mind, to conceive of eternity as a cycle and the eternal purpose as an endless circle, a cycle of eternity in some way synchronized with the transient material cycles of time. As regards the sectors of time connected with, and forming a part of, the cycle of eternity, we are forced to recognize that such temporary epochs are born, live, and die just as the temporary beings of time are born, live, and die. Most human beings die because, having failed to achieve the spirit level of Adjuster fusion, the metamorphosis of death constitutes the only possible procedure whereby they may escape the fetters of time and the bonds of material creation, thereby being enabled to strike spiritual step with the progressive procession of eternity. Having survived the trial life of time and material existence, it becomes possible for you to continue on in touch with, even as a part of, eternity, swinging on forever with the worlds of space around the circle of the eternal ages.

The sectors of time are like the flashes of personality in temporal form; they appear for a season, and then they are lost to human sight, only to reappear as new actors and continuing factors in the higher life of the endless swing around the eternal circle. Eternity can hardly be conceived as a straightaway drive, in view of our belief in a delimited universe moving over a vast, elongated circle around the central dwelling place of the Universal Father.

Frankly, eternity is incomprehensible to the finite mind of time. You simply cannot grasp it; you cannot comprehend it. I do not completely visualize it, and even if I did, it would be impossible for me to convey my concept to the human mind. Nevertheless, I have done my best to portray something of our viewpoint, to tell you somewhat of our understanding of things eternal. I am endeavoring to aid you in the crystallization of your thoughts about these values which are of infinite nature and eternal import.

There is in the mind of God a plan which embraces every creature of all his vast domains, and this plan is an eternal purpose of boundless opportunity, unlimited progress, and endless life. And the infinite treasures of such a matchless career are yours for the striving!

The goal of eternity is ahead! The adventure of divinity attainment lies before you! The race for perfection is on! whosoever will may enter, and certain victory will crown the efforts of every human being who will run the race of faith and trust, depending every step of the way on the leading of the indwelling Adjuster and on the guidance of that good spirit of the Universe Son, which so freely has been poured out upon all flesh.

[Presented by a Mighty Messenger temporarily attached to the Supreme Council of Nebadon and assigned to this mission by Gabriel of Salvington.]

EIGHT

THE LIFE CARRIERS

(PAPER 36)

LIFE does not originate spontaneously. Life is constructed according to plans formulated by the (unrevealed) Architects of Being and appears on the inhabited planets either by direct importation or as a result of the operations of the Life Carriers of the local universes. These carriers of life are among the most interesting and versatile of the diverse family of universe Sons. They are intrusted with designing and carrying creature life to the planetary spheres. And after planting this life on such new worlds, they remain there for long periods to foster its development.

1. Origin and Nature of Life Carriers

Though the Life Carriers belong to the family of divine sonship, they are a peculiar and distinct type of universe Sons, being the only group of intelligent life in a local universe in whose creation the rulers of a superuniverse participate. The Life Carriers are the offspring of three pre-existent personalities: the Creator Son, the Universe Mother Spirit, and, by designation, one of the three Ancients of Days presiding over the destinies of the superuniverse concerned. These Ancients of Days, who alone can decree the extinction of intelligent life, participate in the creation of the Life Carriers, who are intrusted with establishing physical life on the evolving worlds.

In the universe of Nebadon we have on record the creation of one hundred million Life Carriers. This efficient corps of life disseminators is not a truly self-governing group. They are directed by the life-determining trio, consisting of Gabriel, the Father Melchizedek, and Nambia, the original and first-born Life Carrier of Nebadon. But in all phases of their divisional administration they are self-governing.

Life Carriers are graded into three grand divisions: The first division is the senior Life Carriers, the second, assistants, and the third, custodians. The primary division is subdivided into twelve groups of specialists in the various forms of life manifestation. The segregation of these three divisions was effected by the Melchizedeks, who conducted tests for such purposes on the Life Carriers' headquarters sphere. The Melchizedeks have ever since been closely associated with the Life Carriers and always accompany them when they go forth to establish life on a new planet.

When an evolutionary planet is finally settled in light and life, the Life Carriers are organized into the higher deliberative bodies of advisory capacity to assist in the further administration and development of the world and its glorified beings. In the later and settled ages of an evolving universe these Life Carriers are intrusted with many new duties.

2. THE LIFE CARRIER WORLDS

The Melchizedeks have the general oversight of the fourth group of seven primary spheres in the Salvington circuit. These worlds of the Life Carriers are designated as follows:

1. The Life Carrier headquarters.
2. The life-planning sphere.
3. The life-conservation sphere.
4. The sphere of life evolution.
5. The sphere of life associated with mind.
6. The sphere of mind and spirit in living beings.
7. The sphere of unrevealed life.

Each of these primary spheres is surrounded by six satellites, on which the special phases of all the Life Carrier activities in the universe are centered.

World Number One, the headquarters sphere, together with its six tributary satellites, is devoted to the study of universal life, life in all of its known phases of manifestation. Here is located the college of life planning, wherein function teachers and advisers from Uversa and Havona, even from Paradise. And I am permitted to reveal that the seven central emplacements of the adjutant mind-spirits are situated on this world of the Life Carriers.

The number ten—the decimal system—is inherent in the physical universe but not in the spiritual. The domain of life is characterized by three, seven, and twelve or by multiples and combinations of these basic numbers. There are three primal and essentially different life plans, after the order of the three Paradise Sources and Centers, and in the universe of Nebadon these three basic forms of life are segregated on three different types of planets. There were, originally, twelve distinct and divine concepts of transmissible life. This number twelve, with its subdivisions and multiples, runs throughout all basic life patterns of all seven superuniverses. There are also seven architectural types of life design, fundamental arrangements of the reproducing configurations of living matter. The Orvonton life patterns are configured as twelve inheritance carriers. The differing orders of will creatures are configured as 12, 24, 48, 96, 192, 384, and 768. On Urantia there are forty-eight units of pattern control—trait determiners—in the sex cells of human reproduction.

The Second World is the life-designing sphere; here all new modes of life organization are worked out. While the original life designs are provided by the Creator Son, the actual outworking of these plans is intrusted to the Life Carriers and their associates. When the general life plans for a new world have been formulated, they are transmitted to the headquarters sphere, where they are minutely scrutinized by the supreme council of the senior Life Carriers in collaboration with a corps of consulting Melchizedeks. If the plans are a departure from previously accepted formulas, they must be passed upon, and endorsed by, the Creator Son. The chief of Melchizedeks often represents the Creator Son in these deliberations.

Planetary life, therefore, while similar in some respects, differs in many ways on each evolutionary world. Even in a uniform life series in a single family of worlds, life is not exactly the same on any two planets; there is always a planetary type, for the Life Carriers work constantly in an effort to improve the vital formulas committed to their keeping.

There are over one million fundamental or cosmic chemical formulas which constitute the parent patterns and the numerous basic functional variations of life manifestations. Satellite number one of the life-planning sphere is the realm of the universe physicists and electrochemists who serve as technical assistants to the Life Carriers in the work of capturing, organizing, and manipulating the essential units of energy which are employed in building up the material vehicles of life transmission, the so-called germ plasm.

The planetary life-planning laboratories are situated on the second satellite of this world number two. In these laboratories the Life Carriers and all their associates collaborate with the Melchizedeks in the effort to modify and possibly improve the life designed for implantation on the *decimal planets* of Nebadon. The life now evolving on Urantia was planned and partially worked out on this very world, for Urantia is a decimal planet, a life-experiment world. On one world in each ten a greater variance in the standard life designs is permitted than on the other (nonexperimental) worlds.

World Number Three is devoted to the conservation of life. Here various modes of life protection and preservation are studied and developed by the assistants and custodians of the Life Carrier corps. The life plans for every new world always provide for the early establishment of the life-conservation commission, consisting of custodian specialists in the expert manipulation of the basic life patterns. On Urantia there were twenty-four such custodian commissioners, two for each fundamental or parent pattern of the architectural organization of the life material. On planets such as yours the highest form of life is reproduced by a life-carrying bundle which possesses twenty-four pattern units. (And since the intellectual life grows out of, and upon the foundation of, the physical, there come into existence the four and twenty basic orders of psychic organization.)

Sphere Number Four and its tributary satellites are devoted to the study of the evolution of creature life in general and to the evolutionary antecedents of any one life level in particular. The original life plasm of an evolutionary world must contain the full potential for all future developmental variations and for all subsequent evolutionary changes and modifications. The provision for such far-reaching projects of life metamorphosis may require the appearance of many apparently useless forms of animal and vegetable life. Such by-products of planetary evolution, foreseen or unforeseen, appear upon the stage of action only to disappear, but in and through all this long process there runs the thread of the wise and intelligent formulations of the original designers of the planetary life plan and species scheme. The manifold by-products of biologic evolution are all essential to the final and full function of the higher intelligent forms of life, notwithstanding that great outward disharmony may prevail from time to time in the long upward struggle of the higher creatures to effect the mastery of the lower forms of life, many of which are sometimes so antagonistic to the peace and comfort of the evolving will creatures.

Number Five World is concerned wholly with life associated with mind. Each of its satellites is devoted to the study of a single phase of creature mind correlated with creature life. Mind such as man comprehends is an endowment of the seven adjutant mind-spirits superimposed on the nonteachable or mechanical levels of mind by the agencies of the Infinite Spirit. The life patterns are variously responsive to these adjutants and to the different spirit ministries operating throughout the universes of time and space. The capacity of material creatures to effect spirit response is entirely dependent on the associated mind endowment, which, in turn, has directionized the course of the biologic evolution of these same mortal creatures.

World Number Six is dedicated to the correlation of mind with spirit as they are associated with living forms and organisms. This world and its six tributaries embrace the schools of creature co-ordination, wherein teachers from both the central universe and the superuniverse collaborate with the Nebadon instructors in presenting the highest levels of creature attainment in time and space.

The Seventh Sphere of the Life Carriers is dedicated to the unrevealed domains of evolutionary creature life as it is related to the cosmic philosophy of the expanding factualization of the Supreme Being.

3. LIFE TRANSPLANTATION

Life does not spontaneously appear in the universes; the Life Carriers must initiate it on the barren planets. They are the carriers, disseminators, and guardians of life as it appears on the evolutionary worlds of space. All life of the order and forms known on Urantia arises with these Sons, though not all forms of planetary life are existent on Urantia.

The corps of Life Carriers commissioned to plant life upon a new world usually consists of one hundred senior carriers, one hundred assistants, and one thousand custodians. The Life Carriers often carry actual life plasm to a new world, but not always. They sometimes organize the life patterns after arriving on the planet of assignment in accordance with formulas previously approved for a new adventure in life establishment. Such was the origin of the planetary life of Urantia.

When, in accordance with approved formulas, the physical patterns have been provided, then do the Life Carriers catalyze this lifeless material, imparting through their persons the vital spirit spark; and forthwith do the inert patterns become living matter.

The vital spark—the mystery of life—is bestowed through the Life Carriers, not by them. They do indeed supervise such transactions, they formulate the life plasm itself, but it is the Universe Mother Spirit who supplies the essential factor of the living plasm. From the Creative Daughter of the Infinite Spirit comes that energy spark which enlivens the body and presages the mind.

In the bestowal of life the Life Carriers transmit nothing of their personal natures, not even on those spheres where new orders of life are projected. At such times they simply initiate and transmit the spark of life, start the required revolutions of matter in accordance with the physical, chemical, and electrical specifications of the ordained plans and patterns. Life Carriers are living catalytic presences which agitate, organize, and vitalize the otherwise inert elements of the material order of existence.

The Life Carriers of a planetary corps are given a certain period in which to establish life on a new world, approximately one-half million years of the time of that planet. At the termination of this period, indicated by certain developmental attainments of the planetary life, they cease implantation efforts, and they may not subsequently add anything new or supplemental to the life of that planet.

During the ages intervening between life establishment and the emergence of human creatures of moral status, the Life Carriers are permitted to manipulate the life environment and otherwise favorably directionize the course of biologic evolution. And this they do for long periods of time.

When the Life Carriers operating on a new world have once succeeded in producing a being with will, with the power of moral decision and spiritual choice, then and there their work terminates— they are through; they may manipulate the evolving life no further. From this point forward the evolution of living things must proceed in accordance with the endowment of the inherent nature and tendencies which have already been imparted to, and established in, the planetary life formulas and patterns. The Life Carriers are not permitted to experiment or to interfere with will; they are not allowed to dominate or arbitrarily influence moral creatures.

Upon the arrival of a Planetary Prince they prepare to leave, though two of the senior carriers and twelve custodians may volunteer, by taking temporary renunciation vows, to remain indefinitely on the planet as advisers in the matter of the further development and conservation of the life plasm. Two such Sons and their twelve associates are now serving on Urantia.

4. MELCHIZEDEK LIFE CARRIERS

In every local system of inhabited worlds throughout Nebadon there is a single sphere whereon the Melchizedeks have functioned as life carriers. These abodes are known as the system *midsonite* worlds, and on each of them a materially modified Melchizedek Son has mated with a selected Daughter of the material order of sonship. The Mother Eves of such midsonite worlds are dispatched from the system headquarters of jurisdiction, having been chosen by the designated Melchizedek life carrier from among the numerous volunteers who respond to the call of the System Sovereign addressed to the Material Daughters of his sphere.

The progeny of a Melchizedek life carrier and a Material Daughter are known as *midsoniters*. The Melchizedek father of such a race of supernal creatures eventually leaves the planet of his unique life function, and the Mother Eve of this special order of universe beings also departs upon the appearance of the seventh generation of planetary offspring. The direction of such a world then devolves upon her eldest son.

The midsonite creatures live and function as reproducing beings on their magnificent worlds until they are one thousand standard years of age; whereupon they are translated by seraphic transport. Midsoniters are nonreproducing beings thereafter because the technique of dematerialization which they pass through in preparation for enseraphiming forever deprives them of reproductive prerogatives.

The present status of these beings can hardly be reckoned as either mortal or immortal, neither can they be definitely classified as human or divine. These creatures are not Adjuster indwelt, hence hardly immortal. But neither do they seem to be mortal; no midsoniter has experienced death. All midsoniters ever born in Nebadon are alive today, functioning on their native worlds, on some intervening sphere, or on the Salvington midsonite sphere in the finaliters' group of worlds.

The Salvington Worlds of the Finaliters. The Melchizedek life carriers, as well as the associated Mother Eves, go from the system midsonite spheres to the finaliters' worlds of the Salvington circuit, where their offspring are also destined to forgather.

It should be explained in this connection that the fifth group of seven primary worlds in the Salvington circuit are the Nebadon worlds of the finaliters. The children of the Melchizedek life carriers and the Material Daughters are domiciled on the seventh world of the finaliters, the Salvington midsonite sphere.

The satellites of the seven primary worlds of the finaliters are the rendezvous of the personalities of the super- and central universes who may be executing assignments in Nebadon. While the ascending mortals go about freely on all of the cultural worlds and training spheres of the 490 worlds comprising the Melchizedek University, there are certain special schools and numerous restricted zones which they are not permitted to enter. This is especially true of the forty-nine spheres under the jurisdiction of the finaliters.

The purpose of the midsonite creatures is not at present known, but it would appear that these personalities are forgathering on the seventh finaliter world in preparation for some future eventuality in universe evolution. Our inquiries concerning the midsonite races are always referred to the finaliters, and always do the finaliters decline to discuss the destiny of their wards. Regardless of our uncertainty as to the future of the midsoniters, we do know that every local universe in Orvonton harbors such an accumulating corps of these mysterious beings. It is the belief of the Melchizedek life carriers that their midsonite children will some day be endowed with the transcendental and eternal spirit of absonity by God the Ultimate.

5. THE SEVEN ADJUTANT MIND-SPIRITS

It is the presence of the seven adjutant mind-spirits on the primitive worlds that conditions the course of organic evolution; that explains why evolution is purposeful and not accidental. These adjutants represent that function of the mind ministry of the Infinite Spirit which is extended to the lower orders of intelligent life through the operations of a local universe Mother Spirit. The adjutants are the children of the Universe Mother Spirit and constitute her personal ministry to the material minds of the realms. Wherever and whenever such mind is manifest, these spirits are variously functioning.

The seven adjutant mind-spirits are called by names which are the equivalents of the following designations: intuition, understanding, courage, knowledge, counsel, worship, and wisdom. These mind-spirits send forth their influence into all the inhabited worlds as a differential urge, each seeking receptivity capacity for manifestation quite apart from the degree to which its fellows may find reception and opportunity for function.

The central lodgments of the adjutant spirits on the Life Carrier headquarters world indicate to the Life Carrier supervisors the extent and quality of the mind function of the adjutants on any world and in any given living organism of intellect status. These life-mind emplacements are perfect indicators of living mind function for the first five adjutants. But with regard to the sixth and seventh adjutant spirits—worship and wisdom—these central lodgments record only a qualitative function. The quantitative activity of the adjutant of worship and the adjutant of wisdom is registered in the immediate presence of the Divine Minister on Salvington, being a personal experience of the Universe Mother Spirit.

The seven adjutant mind-spirits always accompany the Life Carriers to a new planet, but they should not be regarded as entities; they are more like circuits. The spirits of the seven universe adjutants do not function as personalities apart from the universe presence of the Divine Minister; they are in fact a level of consciousness of the Divine Minister and are always subordinate to the action and presence of their creative mother.

We are handicapped for words adequately to designate these seven adjutant mind-spirits. They are ministers of the lower levels of experiential mind, and they may be described, in the order of evolutionary attainment, as follows:

1. *The spirit of intuition*—quick perception, the primitive physical and inherent reflex instincts, the directional and other self-preservative endowments of all mind creations; the only one of the adjutants to function so largely in the lower orders of animal life and the only one to make extensive functional contact with the non-teachable levels of mechanical mind.

2. *The spirit of understanding*—the impulse of co-ordination, the spontaneous and apparently automatic association of ideas. This is the gift of the co-ordination of acquired knowledge, the phenomenon of quick reasoning, rapid judgment, and prompt decision.

3. *The spirit of courage*—the fidelity endowment—in personal beings, the basis of character acquirement and the intellectual root of moral stamina and spiritual bravery. When enlightened by facts and inspired by truth, this becomes the secret of the urge of evolutionary ascension by the channels of intelligent and conscientious self-direction.

4. *The spirit of knowledge*—the curiosity-mother of adventure and discovery, the scientific spirit; the guide and faithful associate of the spirits of courage and counsel; the urge to direct the endowments of courage into useful and progressive paths of growth.

5. *The spirit of counsel*—the social urge, the endowment of species cooperation; the ability of will creatures to harmonize with their fellows; the origin of the gregarious instinct among the more lowly creatures.

6. *The spirit of worship*—the religious impulse, the first differential urge separating mind creatures into the two basic classes of mortal existence. The spirit of worship forever distinguishes the animal of its association from the soulless creatures of mind endowment. Worship is the badge of spiritual-ascension candidacy.

7. *The spirit of wisdom*—the inherent tendency of all moral creatures towards orderly and progressive evolutionary advancement. This is the highest of the adjutants, the spirit co-ordinator and articulator of the work of all the others. This spirit is the secret of that inborn urge of mind creatures which initiates and maintains the practical and effective program of the ascending scale of existence; that gift of living things which accounts for their inexplicable ability to survive and, in survival, to utilize the co-ordination of all their past experience and present opportunities for the acquisition of all of everything that all of the other six mental ministers can mobilize in the mind of the organism concerned. Wisdom is the acme of intellectual performance. Wisdom is the goal of a purely mental and moral existence.

The adjutant mind-spirits experientially grow, but they never become personal. They evolve in function, and the function of the first five in the animal orders is to a certain extent essential to the function of all seven as human intellect. This animal relationship makes the adjutants more practically effective as human mind; hence animals are to a certain extent indispensable to man's intellectual as well as to his physical evolution.

These mind-adjutants of a local universe Mother Spirit are related to creature life of intelligence status much as the power centers and physical controllers are related to the nonliving forces of the universe. They perform invaluable service in the mind circuits on the inhabited worlds and are effective collaborators with the Master Physical Controllers, who also serve as controllers and directors of the preadjutant mind levels, the levels of nonteachable or mechanical mind.

Living mind, prior to the appearance of capacity to learn from experience, is the ministry domain of the Master Physical Controllers. Creature mind, before acquiring the ability to recognize divinity and worship Deity, is the exclusive domain of the adjutant spirits. With the appearance of the spiritual response of the creature intellect, such created minds at once become superminded, being instantly encircuited in the spirit cycles of the local universe Mother Spirit.

The adjutant mind-spirits are in no manner directly related to the diverse and highly spiritual function of the spirit of the personal presence of the Divine Minister, the Holy Spirit of the inhabited worlds; but they are functionally antecedent to, and preparatory for, the appearance of this very spirit in evolutionary man. The adjutants afford the Universe Mother Spirit a varied contact with, and control over, the material living creatures of a local universe, but they do not repercuss in the Supreme Being when acting on prepersonality levels.

Nonspiritual mind is either a spirit-energy manifestation or a physical-energy phenomenon. Even human mind, personal mind, has no survival qualities apart from spirit identification. Mind is a divinity bestowal, but it is not immortal when it functions without spirit insight, and when it is devoid of the ability to worship and crave survival.

6. LIVING FORCES

Life is both mechanistic and vitalistic—material and spiritual. Ever will Urantia physicists and chemists progress in their understanding of the protoplasmic forms of vegetable and animal life, but never will they be able to produce living organisms. Life is something different from all energy manifestations; even the material life of physical creatures is not inherent in matter.

Things material may enjoy an independent existence, but life springs only from life. Mind can be derived only from pre-existent mind. Spirit takes origin only from spirit ancestors. The creature may produce the forms of life, but only a creator personality or a creative force can supply the activating living spark.

Life Carriers can organize the material forms, or physical patterns, of living beings, but the Spirit provides the initial spark of life and bestows the endowment of mind. Even the living forms of experimental life which the Life Carriers organize on their Salvington worlds are always devoid of reproductive powers. When the life formulas and the vital patterns are correctly assembled and properly organized, the presence of a Life Carrier is sufficient to initiate life, but all such living organisms are lacking in two essential attributes—mind endowment and reproductive powers. Animal mind and human mind are gifts of the local universe Mother Spirit, functioning through the seven adjutant mind-spirits, while creature ability to reproduce is the specific and personal impartation of the Universe Spirit to the ancestral life plasm inaugurated by the Life Carriers.

When the Life Carriers have designed the patterns of life, after they have organized the energy systems, there must occur an additional phenomenon; the "breath of life" must be imparted to these lifeless forms. The Sons of God can construct the forms of life, but it is the Spirit of God who really contributes the vital spark. And when the life thus imparted is spent, then again the remaining material body becomes dead matter. When the bestowed life is exhausted, the body returns to the bosom of the material universe from which it was borrowed by the Life Carriers to serve as a transient vehicle for that life endowment which they conveyed to such a visible association of energy-matter.

The life bestowed upon plants and animals by the Life Carriers does not return to the Life Carriers upon the death of plant or animal. The departing life of such a living thing possesses neither identity nor personality; it does not individually survive death. During its existence and the time of its sojourn in the body of matter, it has undergone a change; it has undergone energy evolution and survives only as a part of the cosmic forces of the universe; it does not survive as individual life. The survival of mortal creatures is wholly predicated on the evolvement of an immortal soul within the mortal mind.

We speak of life as "energy" and as "force," but it is really neither. Force-energy is variously gravity responsive; life is not. Pattern is also nonresponsive to gravity, being a configuration of energies that have already fulfilled all gravity-responsive obligations. Life, as such, constitutes the animation of some pattern-configured or otherwise segregated system of energy—material, mindal, or spiritual.

There are some things connected with the elaboration of life on the evolutionary planets which are not altogether clear to us. We fully comprehend the physical organization of the electrochemical formulas of the Life Carriers, but we do not wholly understand the nature and source of the *life-activation spark*. We know that life flows from the Father through the Son and *by* the Spirit. It is more than possible that the Master Spirits are the sevenfold channel of the river of life which is poured out upon all creation. But we do not comprehend the technique whereby the supervising Master Spirit participates in the initial episode of life bestowal on a new planet. The Ancients of Days, we are confident, also have some part in this inauguration of life on a new world, but we are wholly ignorant of the nature thereof. We do know that the Universe Mother Spirit actually vitalizes the lifeless patterns and imparts to such activated plasm the prerogatives of organismal reproduction. We observe that these three are the levels of God the Sevenfold, sometimes designated as the Supreme Creators of time and space; but otherwise we know little more than Urantia mortals—simply that concept is inherent in the Father, expression in the Son, and life realization in the Spirit.

[Indited by a Vorondadek Son stationed on Urantia as an observer and acting in this capacity by request of the Melchizedek Chief of the Supervising Revelatory Corps.]

NINE

PHYSICAL ASPECTS
OF THE LOCAL UNIVERSE

(PAPER 41)

THE characteristic space phenomenon which sets off each local creation from all others is the presence of the Creative Spirit. All Nebadon is certainly pervaded by the space presence of the Divine Minister of Salvington, and such presence just as certainly terminates at the outer borders of our local universe. That which is pervaded by our local universe Mother Spirit *is* Nebadon; that which extends beyond her space presence is outside Nebadon, being the extra-Nebadon space regions of the superuniverse of Orvonton— other local universes.

While the administrative organization of the grand universe discloses a clear-cut division between the governments of the central, super-, and local universes, and while these divisions are astronomically paralleled in the space separation of Havona and the seven superuniverses, no such clear lines of physical demarcation set off the local creations. Even the major and minor sectors of Orvonton are (to us) clearly distinguishable, but it is not so easy to identify the physical boundaries of the local universes. This is because these local creations are administratively organized in accordance with certain *creative* principles governing the segmentation of the total energy charge of a superuniverse, whereas their physical components, the spheres of space—suns, dark islands, planets, etc.—take origin primarily from nebulae, and these make their astronomical appearance in accordance with certain *precreative* (transcendental) plans of the Architects of the Master Universe.

One or more—even many—such nebulae may be encompassed within the domain of a single local universe even as Nebadon was physically assembled out of the stellar and planetary progeny of Andronover and other nebulae. The spheres of Nebadon are of diverse nebular ancestry, but they all had a certain minimum commonness of space motion which was so adjusted by the intelligent efforts of the power directors as to produce our present aggregation of space bodies, which travel along together as a contiguous unit over the orbits of the superuniverse.

Such is the constitution of the local star cloud of Nebadon, which today swings in an increasingly settled orbit about the Sagittarius center of that minor sector of Orvonton to which our local creation belongs.

1. THE NEBADON POWER CENTERS

The spiral and other nebulae, the mother wheels of the spheres of space, are initiated by Paradise force organizers; and following nebular evolution of gravity response, they are superseded in superuniverse function by the power centers and physical controllers, who thereupon assume full responsibility for directing the physical evolution of the ensuing generations of stellar and planetary offspring. This physical supervision of the Nebadon preuniverse was, upon the arrival of our Creator Son, immediately co-ordinated with his plan for universe organization. Within the domain of this Paradise Son of God, the Supreme Power Centers and the Master Physical Controllers collaborated with the later appearing Morontia Power Supervisors and others to produce that vast complex of communication lines, energy circuits, and power lanes which firmly bind the manifold space bodies of Nebadon into one integrated administrative unit.

One hundred Supreme Power Centers of the fourth order are permanently assigned to our local universe. These beings receive the incoming lines of power from the third-order centers of Uversa and relay the down-stepped and modified circuits to the power centers of our constellations and systems. These power centers, in association, function to produce the living system of control and equalization which operates to maintain the balance and distribution of otherwise fluctuating and variable energies. Power centers are not, however, concerned with transient and local energy upheavals, such as sun spots and system electric disturbances; light and electricity are not the basic energies of space; they are secondary and subsidiary manifestations.

The one hundred local universe centers are stationed on Salvington, where they function at the exact energy center of that sphere. Architectural spheres, such as Salvington, Edentia, and Jerusem, are lighted, heated, and energized by methods which make them quite independent of the suns of space. These spheres were constructed— made to order—by the power centers and physical controllers and were designed to exert a powerful influence over energy distribution. Basing their activities on such focal points of energy control, the power centers, by their living presences, directionize and channelize the physical energies of space. And these energy circuits are basic to all physical-material and morontia-spiritual phenomena.

Ten Supreme Power Centers of the fifth order are assigned to each of Nebadon's primary subdivisions, the one hundred constellations. In Norlatiadek, your constellation, they are not stationed on the headquarters sphere but are situated at the center of the enormous stellar system which constitutes the physical core of the constellation. On Edentia there are ten associated mechanical controllers and ten frandalanks who are in perfect and constant liaison with the near-by power centers.

One Supreme Power Center of the sixth order is stationed at the exact gravity focus of each local system. In the system of Satania the assigned power center occupies a dark island of space located at the astronomic center of the system. Many of these dark islands are vast dynamos which mobilize and directionize certain space-energies, and these natural circumstances are effectively utilized by the Satania Power Center, whose living mass functions as a liaison with the higher centers, directing the streams of more materialized power to the Master Physical Controllers on the evolutionary planets of space.

2. The Satania Physical Controllers

While the Master Physical Controllers serve with the power centers throughout the grand universe, their functions in a local system, such as Satania, are more easy of comprehension. Satania is one of one hundred local systems which make up the administrative organization of the constellation of Norlatiadek, having as immediate neighbors the systems of Sandmatia, Assuntia, Porogia, Sortoria, Rantulia, and Glantonia. The Norlatiadek systems differ in many respects, but all are evolutionary and progressive, very much like Satania.

Satania itself is composed of over seven thousand astronomical groups, or physical systems, few of which had an origin similar to that of your solar system. The astronomic center of Satania is an enormous dark island of space which, with its attendant spheres, is situated not far from the headquarters of the system government.

Except for the presence of the assigned power center, the supervision of the entire physical-energy system of Satania is centered on Jerusem. A Master Physical Controller, stationed on this headquarters sphere, works in coordination with the system power center, serving as liaison chief of the power inspectors headquartered on Jerusem and functioning throughout the local system.

The circuitizing and channelizing of energy is supervised by the five hundred thousand living and intelligent energy manipulators scattered throughout Satania. Through the action of such physical controllers the supervising power centers are in complete and perfect control of a majority of the basic energies of space, including the emanations of highly heated orbs and the dark energy-charged spheres. This group of living entities can mobilize, transform, transmute, manipulate, and transmit nearly all of the physical energies of organized space.

Life has inherent capacity for the mobilization and transmutation of universal energy. You are familiar with the action of vegetable life in transforming the material energy of light into the varied manifestations of the vegetable kingdom. You also know something of the method whereby this vegetative energy can be converted into the phenomena of animal activities, but you know practically nothing of the technique of the power directors and the physical controllers, who are endowed with ability to mobilize, transform, directionize, and concentrate the manifold energies of space.

These beings of the energy realms do not directly concern themselves with energy as a component factor of living creatures, not even with the domain of physiological chemistry. They are sometimes concerned with the physical preliminaries of life, with the elaboration of those energy systems which may serve as the physical vehicles for the living energies of elementary material organisms. In a way the physical controllers are related to the preliving manifestations of material energy as the adjutant mind-spirits are concerned with the prespiritual functions of material mind.

These intelligent creatures of power control and energy direction must adjust their technique on each sphere in accordance with the physical constitution and architecture of that planet. They unfailingly utilize the calculations and deductions of their respective staffs of physicists and other technical advisers regarding the local influence of highly heated suns and other types of supercharged stars. Even the enormous cold and dark giants of space and the swarming clouds of star dust must be reckoned with; all of these material things are concerned in the practical problems of energy manipulation.

The power-energy supervision of the evolutionary inhabited worlds is the responsibility of the Master Physical Controllers, but these beings are not responsible for all energy misbehavior on Urantia. There are a number of reasons for such disturbances, some of which are beyond the domain and control of the physical custodians. Urantia is in the lines of tremendous energies, a small planet in the circuit of enormous masses, and the local controllers sometimes employ enormous numbers of their order in an effort to equalize these lines of energy. They do fairly well with regard to the physical circuits of Satania but have trouble insulating against the powerful Norlatiadek currents.

3. Our Starry Associates

There are upward of two thousand brilliant suns pouring forth light and energy in Satania, and your own sun is an average blazing orb. Of the thirty suns nearest yours, only three are brighter. The

Universe Power Directors initiate the specialized currents of energy which play between the individual stars and their respective systems. These solar furnaces, together with the dark giants of space, serve the power centers and physical controllers as way stations for the effective concentrating and directionizing of the energy circuits of the material creations.

The suns of Nebadon are not unlike those of other universes. The material composition of all suns, dark islands, planets, and satellites, even meteors, is quite identical. These suns have an average diameter of about one million miles, that of your own solar orb being slightly less. The largest star in the universe, the stellar cloud Antares, is four hundred and fifty times the diameter of your sun and is sixty million times its volume. But there is abundant space to accommodate all of these enormous suns. They have just as much comparative elbow room in space as one dozen oranges would have if they were circulating about throughout the interior of Urantia, and were the planet a hollow globe.

When suns that are too large are thrown off a nebular mother wheel, they soon break up or form double stars. All suns are originally truly gaseous, though they may later transiently exist in a semiliquid state. When your sun attained this quasi-liquid state of supergas pressure, it was not sufficiently large to split equatorially, this being one type of double star formation.

When less than one tenth the size of your sun, these fiery spheres rapidly contract, condense, and cool. When upwards of thirty times its size—rather thirty times the gross content of actual material—suns readily split into two separate bodies, either becoming the centers of new systems or else remaining in each other's gravity grasp and revolving about a common center as one type of double star.

The most recent of the major cosmic eruptions in Orvonton was the extraordinary double star explosion, the light of which reached Urantia in A.D.1572. This conflagration was so intense that the explosion was clearly visible in broad daylight.

Not all stars are solid, but many of the older ones are. Some of the reddish, faintly glimmering stars have acquired a density at the center of their enormous masses which would be expressed by saying that one cubic inch of such a star, if on Urantia, would weigh six thousand pounds. The enormous pressure, accompanied by loss of heat and circulating energy, has resulted in bringing the orbits of the basic material units closer and closer together until they now closely approach the status of electronic condensation. This process of cooling and contraction may continue to the limiting and critical explosion point of ultimatonic condensation.

Most of the giant suns are relatively young; most of the dwarf stars are old, but not all. The collisional dwarfs may be very young and may glow with an intense white light, never having known an initial red stage of youthful shining. Both very young and very old suns usually shine with a reddish glow. The yellow tinge indicates moderate youth or approaching old age, but the brilliant white light signifies robust and extended adult life.

While all adolescent suns do not pass through a pulsating stage, at least not visibly, when looking out into space you may observe many of these younger stars whose gigantic respiratory heaves require from two to seven days to complete a cycle. Your own sun still carries a diminishing legacy of the mighty upswellings of its younger days, but the period has lengthened from the former three and one-half day pulsations to the present eleven and one-half year sunspot cycles.

Stellar variables have numerous origins. In some double stars the tides caused by rapidly changing distances as the two bodies swing around their orbits also occasion periodic fluctuations of light. These gravity variations produce regular and recurrent flares, just as the capture of meteors by the accretion of energy-material at the surface would result in a comparatively sudden flash of light which would speedily recede to normal brightness for that sun. Sometimes a sun will capture a stream of meteors in a line of lessened gravity opposition, and occasionally collisions cause stellar flare-ups, but the majority of such phenomena are wholly due to internal fluctuations.

In one group of variable stars the period of light fluctuation is directly dependent on luminosity, and knowledge of this fact enables astronomers to utilize such suns as universe lighthouses or accurate measuring points for the further exploration of distant star clusters. By this technique it is possible to measure stellar distances most precisely up to more than one million light-years. Better methods of space measurement and improved telescopic technique will some-time more fully disclose the ten grand divisions of the superuniverse of Orvonton; you will at least recognize eight of these immense sec-tors as enormous and fairly symmetrical star clusters.

4. Sun Density

The mass of your sun is slightly greater than the estimate of your physicists, who have reckoned it as about two octillion (2 x 1027) tons. It now exists about halfway between the most dense and the most diffuse stars, having about one and one-half times the density of water. But your sun is neither a liquid nor a solid—it is gaseous—and this is true notwithstanding the difficulty of explaining how gaseous matter can attain this and even much greater densities.

Gaseous, liquid, and solid states are matters of atomic-molec-ular relationships, but density is a relationship of space and mass. Density varies directly with the quantity of mass in space and in-versely with the amount of space in mass, the space between the central cores of matter and the particles which whirl around these centers as well as the space within such material particles.

Cooling stars can be physically gaseous and tremendously dense at the same time. You are not familiar with the solar *super-gases*, but these and other unusual forms of matter explain how even nonsolid suns can attain a density equal to iron—about the same as Urantia—and yet be in a highly heated gaseous state and con-tinue to function as suns. The atoms in these dense supergases are exceptionally small; they contain few electrons. Such suns have also largely lost their free ultimatonic stores of energy.

One of your near-by suns, which started life with about the same mass as yours, has now contracted almost to the size of Uran-tia, having become forty thousand times as dense as your sun. The weight of this hot-cold gaseous-solid is about one ton per cubic inch. And still this sun shines with a faint reddish glow, the senile glimmer of a dying monarch of light.

Most of the suns, however, are not so dense. One of your nearer neighbors has a density exactly equal to that of your atmosphere at sea level. If you were in the interior of this sun, you would be unable to discern anything. And temperature permitting, you could penetrate the majority of the suns which twinkle in the night sky and notice no more matter than you perceive in the air of your earthly living rooms.

The massive sun of Veluntia, one of the largest in Orvonton, has a density only one one-thousandth that of Urantia's atmosphere. Were it in composition similar to your atmosphere and not superheated, it would be such a vacuum that human beings would speedily suffocate if they were in or on it.

Another of the Orvonton giants now has a surface temperature a trifle under three thousand degrees. Its diameter is over three hundred million miles—ample room to accommodate your sun and the present orbit of the earth. And yet, for all this enormous size, over forty million times that of your sun, its mass is only about thirty times greater. These enormous suns have an extending fringe that reaches almost from one to the other.

5. SOLAR RADIATION

That the suns of space are not very dense is proved by the steady streams of escaping light-energies. Too great a density would retain light by opacity until the light-energy pressure reached the explosion point. There is a tremendous light or gas pressure within a sun to cause it to shoot forth such a stream of energy as to penetrate space for millions upon millions of miles to energize, light, and heat the distant planets. Fifteen feet of surface of the density of Urantia would effectually prevent the escape of all X rays and light-energies from a sun until the rising internal pressure of accumulating energies resulting from atomic dismemberment overcame gravity with a tremendous outward explosion.

Light, in the presence of the propulsive gases, is highly explosive when confined at high temperatures by opaque retaining walls. Light is real. As you value energy and power on your world, sunlight would be economical at a million dollars a pound.

The interior of your sun is a vast X-ray generator. The suns are supported from within by the incessant bombardment of these mighty emanations.

It requires more than one-half million years for an X-ray-stimulated electron to work its way from the very center of an average sun up to the solar surface, whence it starts out on its space adventure, maybe to warm an inhabited planet, to be captured by a meteor, to participate in the birth of an atom, to be attracted by a highly charged dark island of space, or to find its space flight terminated by a final plunge into the surface of a sun similar to the one of its origin.

The X rays of a sun's interior charge the highly heated and agitated electrons with sufficient energy to carry them out through space, past the hosts of detaining influences of intervening matter and, in spite of divergent gravity attractions, on to the distant spheres of the remote systems. The great energy of velocity required to escape the gravity clutch of a sun is sufficient to insure that the sunbeam will travel on with unabated velocity until it encounters considerable masses of matter; whereupon it is quickly transformed into heat with the liberation of other energies.

Energy, whether as light or in other forms, in its flight through space moves straight forward. The actual particles of material existence traverse space like a fusillade. They go in a straight and unbroken line or procession except as they are acted on by superior forces, and except as they ever obey the linear-gravity pull inherent in material mass and the circular-gravity presence of the Isle of Paradise.

Solar energy may seem to be propelled in waves, but that is due to the action of coexistent and diverse influences. A given form of organized energy does not proceed in waves but in direct lines. The presence of a second or a third form of force-energy may cause the stream under observation to *appear* to travel in wavy formation, just as, in a blinding rainstorm accompanied by a heavy wind, the water sometimes appears to fall in sheets or to descend in waves. The raindrops are coming down in a direct line of unbroken procession, but the action of the wind is such as to give the visible appearance of sheets of water and waves of raindrops.

The action of certain secondary and other undiscovered energies present in the space regions of your local universe is such that solar-light emanations appear to execute certain wavy phenomena as well as to be chopped up into infinitesimal portions of definite length and weight. And, practically considered, that is exactly what happens. You can hardly hope to arrive at a better understanding of the behavior of light until such a time as you acquire a clearer concept of the interaction and interrelationship of the various space-forces and solar energies operating in the space regions of Nebadon.

Your present confusion is also due to your incomplete grasp of this problem as it involves the interassociated activities of the personal and nonpersonal control of the master universe—the presences, the performances, and the co-ordination of the Conjoint Actor and the Unqualified Absolute.

6. Calcium—The Wanderer of Space

In deciphering spectral phenomena, it should be remembered that space is not empty; that light, in traversing space, is sometimes slightly modified by the various forms of energy and matter which circulate in all organized space. Some of the lines indicating unknown matter which appear in the spectra of your sun are due to modifications of well-known elements which are floating throughout space in shattered form, the atomic casualties of the fierce encounters of the solar elemental battles. Space is pervaded by these wandering derelicts, especially sodium and calcium.

Calcium is, in fact, the chief element of the matter-permeation of space throughout Orvonton. Our whole superuniverse is sprinkled with minutely pulverized stone. Stone is literally the basic building matter for the planets and spheres of space. The cosmic cloud, the great space blanket, consists for the most part of the modified atoms of calcium. The stone atom is one of the most prevalent and persistent of the elements. It not only endures solar ionization—splitting—but persists in an associative identity even after it has been battered by the destructive X rays and shattered by the high solar temperatures. Calcium possesses an individuality and a longevity excelling all of the more common forms of matter.

As your physicists have suspected, these mutilated remnants of solar calcium literally ride the light beams for varied distances, and thus their widespread dissemination throughout space is tremendously facilitated. The sodium atom, under certain modifications, is also capable of light and energy locomotion. The calcium feat is all the more remarkable since this element has almost twice the mass of sodium. Local space-permeation by calcium is due to the fact that it escapes from the solar photosphere, in modified form, by literally riding the outgoing sunbeams. Of all the solar elements, calcium, notwithstanding its comparative bulk—containing as it does twenty revolving electrons—is the most successful in escaping from the solar interior to the realms of space. This explains why there is a calcium layer, a gaseous stone surface, on the sun six thousand miles thick; and this despite the fact that nineteen lighter elements, and numerous heavier ones, are underneath.

Calcium is an active and versatile element at solar temperatures. The stone atom has two agile and loosely attached electrons in the two outer electronic circuits, which are very close together. Early in the atomic struggle it loses its outer electron; whereupon it engages in a masterful act of juggling the nineteenth electron back and forth between the nineteenth and twentieth circuits of electronic revolution. By tossing this nineteenth electron back and forth between its own orbit and that of its lost companion more than twenty-five thousand times a second, a mutilated stone atom is able partially to defy gravity and thus successfully to ride the emerging streams of light and energy, the sunbeams, to liberty and adventure. This calcium atom moves outward by alternate jerks of forward propulsion, grasping and letting go the sunbeam about twenty-five thousand times each second. And this is why stone is the chief component of the worlds of space. Calcium is the most expert solar-prison escaper.

The agility of this acrobatic calcium electron is indicated by the fact that, when tossed by the temperature-X-ray solar forces to the circle of the higher orbit, it only remains in that orbit for about one one-millionth of a second; but before the electric-gravity power of the atomic nucleus pulls it back into its old orbit, it is able to complete one million revolutions about the atomic center.

Your sun has parted with an enormous quantity of its calcium, having lost tremendous amounts during the times of its convulsive eruptions in connection with the formation of the solar system. Much of the solar calcium is now in the outer crust of the sun.

It should be remembered that spectral analyses show only sunsurface compositions. For example: Solar spectra exhibit many iron lines, but iron is not the chief element in the sun. This phenomenon is almost wholly due to the present temperature of the sun's surface, a little less than 6,000 degrees, this temperature being very favorable to the registry of the iron spectrum.

7. SOURCES OF SOLAR ENERGY

The internal temperature of many of the suns, even your own, is much higher than is commonly believed. In the interior of a sun practically no whole atoms exist; they are all more or less shattered by the intensive X-ray bombardment which is indigenous to such high temperatures. Regardless of what material elements may appear in the outer layers of a sun, those in the interior are rendered very similar by the dissociative action of the disruptive X rays. X ray is the great leveler of atomic existence.

The surface temperature of your sun is almost 6,000 degrees, but it rapidly increases as the interior is penetrated until it attains the unbelievable height of about 35,000,000 degrees in the central regions. (All of these temperatures refer to your Fahrenheit scale.)

All of these phenomena are indicative of enormous energy expenditure, and the sources of solar energy, named in the order of their importance, are:

1. Annihilation of atoms and, eventually, of electrons.

2. Transmutation of elements, including the radioactive group of energies thus liberated.

3. The accumulation and transmission of certain universal space-energies.

4. Space matter and meteors which are incessantly diving into the blazing suns.

5. Solar contraction; the cooling and consequent contraction of a sun yields energy and heat sometimes greater than that supplied by space matter.

6. Gravity action at high temperatures transforms certain circuitized power into radiative energies.

7. Receptive light and other matter which are drawn back into the sun after having left it, together with other energies having extrasolar origin.

There exists a regulating blanket of hot gases (sometimes millions of degrees in temperature) which envelops the suns, and which acts to stabilize heat loss and otherwise prevent hazardous fluctuations of heat dissipation. During the active life of a sun the internal temperature of 35,000,000 degrees remains about the same quite regardless of the progressive fall of the external temperature.

You might try to visualize 35,000,000 degrees of heat, in association with certain gravity pressures, as the electronic boiling point. Under such pressure and at such temperature all atoms are degraded and broken up into their electronic and other ancestral components; even the electrons and other associations of ultimatons may be broken up, but the suns are not able to degrade the ultimatons.

These solar temperatures operate to enormously speed up the ultimatons and the electrons, at least such of the latter as continue to maintain their existence under these conditions. You will realize what high temperature means by way of the acceleration of ultima-tonic and electronic activities when you pause to consider that one drop of ordinary water contains over one billion trillions of atoms. This is the energy of more than one hundred horsepower exerted continuously for two years. The total heat now given out by the solar system sun each second is sufficient to boil all the water in all the oceans on Urantia in just one second of time.

Only those suns which function in the direct channels of the main streams of universe energy can shine on forever. Such solar furnaces blaze on indefinitely, being able to replenish their material losses by the intake of space-force and analogous circulating energy. But stars far removed from these chief channels of recharging are destined to undergo energy depletion—gradually cool off and eventually burn out.

Such dead or dying suns can be rejuvenated by collisional impact or can be recharged by certain nonluminous energy islands of space or through gravity-robbery of near-by smaller suns or systems. The majority of dead suns will experience revivification by these or other evolutionary techniques. Those which are not thus eventually recharged are destined to undergo disruption by mass explosion when the gravity condensation attains the critical level of ultimatonic condensation of energy pressure. Such disappearing suns thus become energy of the rarest form, admirably adapted to energize other more favorably situated suns.

8. SOLAR-ENERGY REACTIONS

In those suns which are encircuited in the space-energy channels, solar energy is liberated by various complex nuclear-reaction chains, the most common of which is the hydrogen-carbon-helium reaction. In this metamorphosis, carbon acts as an energy catalyst since it is in no way actually changed by this process of converting hydrogen into helium. Under certain conditions of high temperature the hydrogen penetrates the carbon nuclei. Since the carbon cannot hold more than four such protons, when this saturation state is attained, it begins to emit protons as fast as new ones arrive. In this reaction the ingoing hydrogen particles come forth as a helium atom.

Reduction of hydrogen content increases the luminosity of a sun. In the suns destined to burn out, the height of luminosity is attained at the point of hydrogen exhaustion. Subsequent to this point, brilliance is maintained by the resultant process of gravity contraction. Eventually, such a star will become a so-called white dwarf, a highly condensed sphere.

In large suns—small circular nebulae—when hydrogen is exhausted and gravity contraction ensues, if such a body is not sufficiently opaque to retain the internal pressure of support for the outer gas regions, then a sudden collapse occurs. The gravity-electric changes give origin to vast quantities of tiny particles devoid of electric potential, and such particles readily escape from the solar interior, thus bringing about the collapse of a gigantic sun within a few days. It was such an emigration of these "runaway particles" that occasioned the collapse of the giant nova of the Andromeda nebula about fifty years ago.[1] This vast stellar body collapsed in forty minutes of Urantia time.

As a rule, the vast extrusion of matter continues to exist about the residual cooling sun as extensive clouds of nebular gases. And all this explains the origin of many types of irregular nebulae, such as the Crab nebula, which had its origin about nine hundred years ago, and which still exhibits the mother sphere as a lone star near the center of this irregular nebular mass.

9. Sun Stability

The larger suns maintain such a gravity control over their electrons that light escapes only with the aid of the powerful X rays.

---

[1] From A.D. 1934, when this paper was received.

These helper rays penetrate all space and are concerned in the maintenance of the basic ultimatonic associations of energy. The great energy losses in the early days of a sun, subsequent to its attainment of maximum temperature—upwards of 35,000,000 degrees—are not so much due to light escape as to ultimatonic leakage. These ultimaton energies escape out into space, to engage in the adventure of electronic association and energy materialization, as a veritable energy blast during adolescent solar times.

Atoms and electrons are subject to gravity. The ultimatons are *not* subject to local gravity, the interplay of material attraction, but they are fully obedient to absolute or Paradise gravity, to the trend, the swing, of the universal and eternal circle of the universe of universes. Ultimatonic energy does not obey the linear or direct gravity attraction of near-by or remote material masses, but it does ever swing true to the circuit of the great ellipse of the far-flung creation.

Your own solar center radiates almost one hundred billion tons of actual matter annually, while the giant suns lose matter at a prodigious rate during their earlier growth, the first billion years. A sun's life becomes stable after the maximum of internal temperature is reached, and the subatomic energies begin to be released. And it is just at this critical point that the larger suns are given to convulsive pulsations.

Sun stability is wholly dependent on the equilibrium between gravity-heat contention—tremendous pressures counterbalanced by unimagined temperatures. The interior gas elasticity of the suns upholds the overlying layers of varied materials, and when gravity and heat are in equilibrium, the weight of the outer materials exactly equals the temperature pressure of the underlying and interior gases. In many of the younger stars continued gravity condensation produces ever-heightening internal temperatures, and as internal heat increases, the interior X-ray pressure of supergas winds becomes so great that, in connection with the centrifugal motion, a sun begins to throw its exterior layers off into space, thus redressing the imbalance between gravity and heat.

Your own sun has long since attained relative equilibrium between its expansion and contraction cycles, those disturbances which produce the gigantic pulsations of many of the younger stars.

Your sun is now passing out of its six billionth year. At the present time it is functioning through the period of greatest economy. It will shine on as of present efficiency for more than twenty-five billion years. It will probably experience a partially efficient period of decline as long as the combined periods of its youth and stabilized function.

10. Origin of Inhabited Worlds

Some of the variable stars, in or near the state of maximum pulsation, are in process of giving origin to subsidiary systems, many of which will eventually be much like your own sun and its revolving planets. Your sun was in just such a state of mighty pulsation when the massive Angona system swung into near approach, and the outer surface of the sun began to erupt veritable streams—continuous sheets—of matter. This kept up with ever-increasing violence until nearest apposition, when the limits of solar cohesion were reached and a vast pinnacle of matter, the ancestor of the solar system, was disgorged. In similar circumstances the closest approach of the attracting body sometimes draws off whole planets, even a quarter or third of a sun. These major extrusions form certain peculiar cloud-bound types of worlds, spheres much like Jupiter and Saturn.

The majority of solar systems, however, had an origin entirely different from yours, and this is true even of those which were produced by gravity-tidal technique. But no matter what technique of world building obtains, gravity always produces the solar system type of creation; that is, a central sun or dark island with planets, satellites, subsatellites, and meteors.

The physical aspects of the individual worlds are largely determined by mode of origin, astronomical situation, and physical environment. Age, size, rate of revolution, and velocity through space are also determining factors. Both the gas-contraction and the solid-accretion worlds are characterized by mountains and, during their earlier life, when not too small, by water and air. The molten-split and collisional worlds are sometimes without extensive mountain ranges.

During the earlier ages of all these new worlds, earthquakes are frequent, and they are all characterized by great physical disturbances; especially is this true of the gas-contraction spheres, the worlds born of the immense nebular rings which are left behind in the wake of the early condensation and contraction of certain individual suns. Planets having a dual origin like Urantia pass through a less violent and stormy youthful career. Even so, your world experienced an early phase of mighty upheavals, characterized by volcanoes, earthquakes, floods, and terrific storms.

Urantia is comparatively isolated on the outskirts of Satania, your solar system, with one exception, being the farthest removed from Jerusem, while Satania itself is next to the outermost system of Norlatiadek, and this constellation is now traversing the outer fringe of Nebadon. You were truly among the least of all creation until Michael's bestowal elevated your planet to a position of honor and great universe interest. Sometimes the last is first, while truly the least becomes greatest.

[Presented by an Archangel in collaboration with the Chief of Nebadon Power Centers.]

TEN

THE CONSTELLATIONS

(PAPER 43)

URANTIA is commonly referred to as 606 of Satania in Norlatiadek of Nebadon, meaning the six hundred sixth inhabited world in the local system of Satania, situated in the constellation of Norlatiadek, one of the one hundred constellations of the local universe of Nebadon. Constellations being the primary divisions of a local universe, their rulers link the local systems of inhabited worlds to the central administration of the local universe on Salvington and by reflectivity to the superadministration of the Ancients of Days on Uversa.

The government of your constellation is situated in a cluster of 771 architectural spheres, the centermost and largest of which is Edentia, the seat of the administration of the Constellation Fathers, the Most Highs of Norlatiadek. Edentia itself is approximately one hundred times as large as your world. The seventy major spheres surrounding Edentia are about ten times the size of Urantia, while the ten satellites which revolve around each of these seventy worlds are about the size of Urantia. These 771 architectural spheres are quite comparable in size to those of other constellations.

Edentia time reckoning and distance measurement are those of Salvington, and like the spheres of the universe capital, the constellation headquarters worlds are fully supplied with all orders of celestial intelligences. In general, these personalities are not very different from those described in connection with the universe administration.

The supervisor seraphim, the third order of local universe angels, are assigned to the service of the constellations. They make their headquarters on the capital spheres and minister extensively to the encircling morontia-training worlds. In Norlatiadek the seventy major spheres, together with the seven hundred minor satellites, are inhabited by the univitatia, the permanent citizens of the constellation. All these architectural worlds are fully administered by the various groups of native life, for the greater part unrevealed but including the efficient spironga and the beautiful spornagia. Being the mid-point in the morontia-training regime, as you might suspect, the morontia life of the constellations is both typical and ideal.

1. THE CONSTELLATION HEADQUARTERS

Edentia abounds in fascinating highlands, extensive elevations of physical matter crowned with morontia life and overspread with spiritual glory, but there are no rugged mountain ranges such as appear on Urantia. There are tens of thousands of sparkling lakes and thousands upon thousands of interconnecting streams, but there are no great oceans nor torrential rivers. Only the highlands are devoid of these surface streams.

The water of Edentia and similar architectural spheres is no different from the water of the evolutionary planets. The water systems of such spheres are both surface and subterranean, and the moisture is in constant circulation. Edentia can be circumnavigated via these various water routes, though the chief channel of transportation is the atmosphere. Spirit beings would naturally travel above the surface of the sphere, while the morontia and material beings make use of material and semimaterial means to negotiate atmospheric passage.

Edentia and its associated worlds have a true atmosphere, the usual three-gas mixture which is characteristic of such architectural creations, and which embodies the two elements of Urantian atmosphere plus that morontia gas suitable for the respiration of morontia creatures. But while this atmosphere is both material and morontial, there are no storms or hurricanes; neither is there summer nor winter. This absence of atmospheric disturbances and of seasonal variation makes it possible to embellish all outdoors on these especially created worlds.

The Edentia highlands are magnificent physical features, and their beauty is enhanced by the endless profusion of life which abounds throughout their length and breadth. Excepting a few rather isolated structures, these highlands contain no work of creature hands. Material and morontial ornamentations are limited to the dwelling areas. The lesser elevations are the sites of special residences and are beautifully embellished with both biologic and morontia art.

Situated on the summit of the seventh highland range are the resurrection halls of Edentia, wherein awaken the ascending mortals of the secondary modified order of ascension. These chambers of creature reassembly are under the supervision of the Melchizedeks. The first of the receiving spheres of Edentia (like the planet Melchizedek near Salvington) also has special resurrection halls, wherein the mortals of the modified orders of ascension are reassembled.

The Melchizedeks also maintain two special colleges on Edentia. One, the emergency school, is devoted to the study of problems growing out of the Satania rebellion. The other, the bestowal school, is dedicated to the mastery of the new problems arising out of the fact that Michael made his final bestowal on one of the worlds of Norlatiadek. This latter college was established almost forty thousand years ago, immediately after the announcement by Michael that Urantia had been selected as the world for his final bestowal.

The sea of glass, the receiving area of Edentia, is near the administrative center and is encircled by the headquarters amphitheater. Surrounding this area are the governing centers for the seventy divisions of constellation affairs. One half of Edentia is divided into seventy triangular sections, whose boundaries converge at the headquarters buildings of their respective sectors. The remainder of this sphere is one vast natural park, the gardens of God.

During your periodic visits to Edentia, though the entire planet is open to your inspection, most of your time will be spent in that administrative triangle whose number corresponds to that of your current residential world. You will always be welcome as an observer in the legislative assemblies.

The morontia area assigned to ascending mortals resident on Edentia is located in the mid-zone of the thirty-fifth triangle adjoining the headquarters of the finaliters, situated in the thirty-sixth triangle. The general headquarters of the univitatia occupies an enormous area in the mid-region of the thirty-fourth triangle immediately adjoining the residential reservation of the morontia citizens. From these arrangements it may be seen that provision is made for the accommodation of at least seventy major divisions of celestial life, and also that each of these seventy triangular areas is correlated with some one of the seventy major spheres of morontia training.

The Edentia sea of glass is one enormous circular crystal about one hundred miles in circumference and about thirty miles in depth. This magnificent crystal serves as the receiving field for all transport seraphim and other beings arriving from points outside the sphere; such a sea of glass greatly facilitates the landing of transport seraphim.

A crystal field on this order is found on almost all architectural worlds; and it serves many purposes aside from its decorative value, being utilized for portraying superuniverse reflectivity to assembled groups and as a factor in the energy-transformation technique for modifying the currents of space and for adapting other incoming physical-energy streams.

2. The Constellation Government

The constellations are the autonomous units of a local universe, each constellation being administered according to its own legislative enactments. When the courts of Nebadon sit in judgment on universe affairs, all internal matters are adjudicated in accordance with the laws prevailing in the constellation concerned. These judicial decrees of Salvington, together with the legislative enactments of the constellations, are executed by the administrators of the local systems.

Constellations thus function as the legislative or lawmaking units, while the local systems serve as the executive or enforcement units. The Salvington government is the supreme judicial and co-ordinating authority.

While the supreme judicial function rests with the central administration of a local universe, there are two subsidiary but major tribunals at the headquarters of each constellation, the Melchizedek council and the court of the Most High.

All judicial problems are first reviewed by the council of the Melchizedeks. Twelve of this order who have had certain requisite experience on the evolutionary planets and on the system headquarters worlds are empowered to review evidence, digest pleas, and formulate provisional verdicts, which are passed on to the court of the Most High, the reigning Constellation Father. The mortal division of this latter tribunal consists of seven judges, all of whom are ascendant mortals. The higher you ascend in the universe, the more certain you are to be judged by those of your own kind.

The constellation legislative body is divided into three groups. The legislative program of a constellation originates in the lower house of ascenders, a group presided over by a finaliter and consisting of one thousand representative mortals. Each system nominates ten members to sit in this deliberative assembly. On Edentia this body is not fully recruited at the present time.

The mid-chamber of legislators is composed of the seraphic hosts and their associates, other children of the local universe Mother Spirit. This group numbers one hundred and is nominated by the supervising personalities who preside over the various activities of such beings as they function within the constellation.

The advisory or highest body of constellation legislators consists of the house of peers—the house of the divine Sons. This corps is chosen by the Most High Fathers and numbers ten. Only Sons of special experience may serve in this upper house. This is the factfinding and timesaving group which very effectively serves both of the lower divisions of the legislative assembly.

The combined council of legislators consists of three members from each of these separate branches of the constellation deliberative assembly and is presided over by the reigning junior Most High. This group sanctions the final form of all enactments and authorizes their promulgation by the broadcasters. The approval of this supreme commission renders legislative enactments the law of the realm; their acts are final. The legislative pronouncements of Edentia constitute the fundamental law of all Norlatiadek.

3. THE MOST HIGHS OF NORLATIADEK

The rulers of the constellations are of the Vorondadek order of local universe sonship. When commissioned to active duty in the universe as constellation rulers or otherwise, these Sons are known as the *Most Highs* since they embody the highest administrative wisdom, coupled with the most farseeing and intelligent loyalty, of all the orders of the Local Universe Sons of God. Their personal integrity and their group loyalty have never been questioned; no disaffection of the Vorondadek Sons has ever occurred in Nebadon.

At least three Vorondadek Sons are commissioned by Gabriel as the Most Highs of each of the Nebadon constellations. The presiding member of this trio is known as the *Constellation Father* and his two associates as the *senior Most High* and the *junior Most High*. A Constellation Father reigns for ten thousand standard years (about 50,000 Urantia years), having previously served as junior associate and as senior associate for equal periods.

The Psalmist knew that Edentia was ruled by three Constellation Fathers and accordingly spoke of their abode in the plural: "There is a river, the streams whereof shall make glad the city of God, the most holy place of the tabernacles of the Most Highs."

Down through the ages there has been great confusion on Urantia regarding the various universe rulers. Many later teachers confused their vague and indefinite tribal deities with the Most High Fathers. Still later, the Hebrews merged all of these celestial rulers into a composite Deity. One teacher understood that the Most Highs were not the Supreme Rulers, for he said, "He who dwells in the secret place of the Most High shall abide under the shadow of the Almighty." In the Urantia records it is very difficult at times to know exactly who is referred to by the term "Most High." But Daniel fully understood these matters. He said, "The Most High rules in the kingdom of men and gives it to whomsoever he will."

The Constellation Fathers are little occupied with the individuals of an inhabited planet, but they are closely associated with those legislative and lawmaking functions of the constellations which so greatly concern every mortal *race* and national *group* of the inhabited worlds.

Although the constellation regime stands between you and the universe administration, as individuals you would ordinarily be little concerned with the constellation government. Your great interest would normally center in the local system, Satania; but temporarily, Urantia is closely related to the constellation rulers because of certain system and planetary conditions growing out of the Lucifer rebellion.

The Edentia Most Highs seized certain phases of planetary authority on the rebellious worlds at the time of the Lucifer secession. They have continued to exercise this power, and the Ancients of Days long since confirmed this assumption of control over these wayward worlds. They will no doubt continue to exercise this assumed jurisdiction as long as Lucifer lives. Much of this authority would ordinarily, in a loyal system, be invested in the System Sovereign.

But there is still another way in which Urantia became peculiarly related to the Most Highs. When Michael, the Creator Son, was on his terminal bestowal mission, since the successor of Lucifer was not in full authority in the local system, all Urantia affairs which concerned the Michael bestowal were immediately supervised by the Most Highs of Norlatiadek.

4. MOUNT ASSEMBLY—THE FAITHFUL OF DAYS

The most holy mount of assembly is the dwelling place of the Faithful of Days, the representative of the Paradise Trinity who functions on Edentia.

This Faithful of Days is a Trinity Son of Paradise and has been present on Edentia as the personal representative of Immanuel since the creation of the headquarters world. Ever the Faithful of Days stands at the right hand of the Constellation Fathers to counsel them, but never does he proffer advice unless it is asked for. The high Sons of Paradise never participate in the conduct of the affairs of a local universe except upon the petition of the acting rulers of such domains. But all that a Union of Days is to a Creator Son, a Faithful of Days is to the Most Highs of a constellation.

The residence of the Edentia Faithful of Days is the constellation center of the Paradise system of extra-universe communication and intelligence. These Trinity Sons, with their staffs of Havona and Paradise personalities, in liaison with the supervising Union of Days, are in direct and constant communication with their order throughout all the universes, even to Havona and Paradise.

The most holy mount is exquisitely beautiful and marvelously appointed, but the actual residence of the Paradise Son is modest in comparison with the central abode of the Most Highs and the surrounding seventy structures comprising the residential unit of the Vorondadek Sons. These appointments are exclusively residential; they are entirely separate from the extensive administrative headquarters buildings wherein the affairs of the constellation are transacted.

The residence of the Faithful of Days on Edentia is located to the north of these residences of the Most Highs and is known as the "mount of Paradise assembly." On this consecrated highland the ascending mortals periodically assemble to hear this Son of Paradise tell of the long and intriguing journey of progressing mortals through the one billion perfection worlds of Havona and on to the indescribable delights of Paradise. And it is at these special gatherings on Mount Assembly that the morontia mortals become more fully acquainted with the various groups of personalities of origin in the central universe.

The traitorous Lucifer, onetime sovereign of Satania, in announcing his claims to increased jurisdiction, sought to displace all superior orders of sonship in the governmental plan of the local universe. He purposed in his heart, saying: "I will exalt my throne above the Sons of God; I will sit upon the mount of assembly in the north; I will be like the Most High."

The one hundred System Sovereigns come periodically to the Edentia conclaves which deliberate on the welfare of the constellation. After the Satania rebellion the archrebels of Jerusem were wont to come up to these Edentia councils just as they had on former occasions. And there was found no way to stop this arrogant effrontery until after the bestowal of Michael on Urantia and his subsequent assumption of unlimited sovereignty throughout all Nebadon. Never, since that day, have these instigators of sin been permitted to sit in the Edentia councils of the loyal System Sovereigns.

That the teachers of olden times knew of these things is shown by the record: "And there was a day when the Sons of God came to present themselves before the Most Highs, and Satan came also and presented himself among them." And this is a statement of fact regardless of the connection in which it chances to appear.

Since the triumph of Christ, all Norlatiadek is being cleansed of sin and rebels. Sometime before Michael's death in the flesh the fallen Lucifer's associate, Satan, sought to attend such an Edentia conclave, but the solidification of sentiment against the archrebels had reached the point where the doors of sympathy were so well-nigh universally closed that there could be found no standing ground for the Satania adversaries. When there exists no open door for the reception of evil, there exists no opportunity for the entertainment of sin. The doors of the hearts of all Edentia closed against Satan; he was unanimously rejected by the assembled System Sovereigns, and it was at this time that the Son of Man "beheld Satan fall as lightning from heaven."

Since the Lucifer rebellion a new structure has been provided near the residence of the Faithful of Days. This temporary edifice is the headquarters of the Most High liaison, who functions in close touch with the Paradise Son as adviser to the constellation government in all matters respecting the policy and attitude of the order of Days toward sin and rebellion.

5. THE EDENTIA FATHERS SINCE THE LUCIFER REBELLION

The rotation of the Most Highs on Edentia was suspended at the time of the Lucifer rebellion. We now have the same rulers who were on duty at that time. We infer that no change in these rulers will be made until Lucifer and his associates are finally disposed of.

The present government of the constellation, however, has been expanded to include twelve Sons of the Vorondadek order. These twelve are as follows:

1. The Constellation Father. The present Most High ruler of Norlatiadek is number 617,318 of the Vorondadek series of Nebadon. He saw service in many constellations throughout our local universe before taking up his Edentia responsibilities.

2. The senior Most High associate.

3. The junior Most High associate.

4. The Most High adviser, the personal representative of Michael since his attainment of the status of a Master Son.

5. The Most High executive, the personal representative of Gabriel stationed on Edentia ever since the Lucifer rebellion.

6. The Most High chief of planetary observers, the director of the Vorondadek observers stationed on the isolated worlds of Satania.

7. The Most High referee, the Vorondadek Son intrusted with the duty of adjusting all difficulties consequential to rebellion within the constellation.

8. The Most High emergency administrator, the Vorondadek Son charged with the task of adapting the emergency enactments of the Norlatiadek legislature to the rebellion-isolated worlds of Satania.

9. The Most High mediator, the Vorondadek Son assigned to harmonize the special bestowal adjustments on Urantia with the routine administration of the constellation. The presence of certain archangel activities and numerous other irregular ministrations on Urantia, together with the special activities of the Brilliant Evening Stars on Jerusem, necessitates the functioning of this Son.

10. The Most High judge-advocate, the head of the emergency tribunal devoted to the adjustment of the special problems of Norlatiadek growing out of the confusion consequent upon the Satania rebellion.

11. The Most High liaison, the Vorondadek Son attached to the Edentia rulers but commissioned as a special counselor with the Faithful of Days regarding the best course to pursue in the management of problems pertaining to rebellion and creature disloyalty.

12. The Most High director, the president of the emergency council of Edentia. All personalities assigned to Norlatiadek because of the Satania upheaval constitute the emergency council, and their presiding officer is a Vorondadek Son of extraordinary experience.

And this takes no account of the numerous Vorondadeks, envoys of Nebadon constellations, and others who are also resident on Edentia.

Ever since the Lucifer rebellion the Edentia Fathers have exercised a special care over Urantia and the other isolated worlds of Satania. Long ago the prophet recognized the controlling hand of the Constellation Fathers in the affairs of nations. "When the Most High divided to the nations their inheritance, when he separated the sons of Adam, he set the bounds of the people."

Every quarantined or isolated world has a Vorondadek Son acting as an observer. He does not participate in planetary administration except when ordered by the Constellation Father to intervene in the affairs of the nations. Actually it is this Most High observer who "rules in the kingdoms of men." Urantia is one of the isolated worlds of Norlatiadek, and a Vorondadek observer has been stationed on the planet ever since the Caligastia betrayal. When Machiventa Melchizedek ministered in semimaterial form on Urantia, he paid respectful homage to the Most High observer then on duty, as it is written, "And Melchizedek, king of Salem, was the priest of the Most High." Melchizedek revealed the relations of this Most High observer to Abraham when he said, "And blessed be the Most High, who has delivered your enemies into your hand."

6. THE GARDENS OF GOD

The system capitals are particularly beautified with material and mineral constructions, while the universe headquarters is more reflective of spiritual glory, but the capitals of the constellations are the acme of morontia activities and living embellishments. On the constellation headquarters worlds living embellishment is more generally utilized, and it is this preponderance of life—botanic artistry—that causes these worlds to be called "the gardens of God."

About one half of Edentia is devoted to the exquisite gardens of the Most Highs, and these gardens are among the most entrancing morontia creations of the local universe. This explains why the extraordinarily beautiful places on the inhabited worlds of Norlatiadek are so often called "the garden of Eden."

Centrally located in this magnificent garden is the worship shrine of the Most Highs. The Psalmist must have known something about these things, for he wrote: "Who shall ascend the hill of the Most Highs? Who shall stand in this holy place? He who has clean hands and a pure heart, who has not lifted up his soul to vanity nor sworn deceitfully." At this shrine the Most Highs, on every tenth day of relaxation, lead all Edentia in the worshipful contemplation of God the Supreme.

The architectural worlds enjoy ten forms of life of the material order. On Urantia there is plant and animal life, but on such a world as Edentia there are ten divisions of the material orders of life. Were you to view these ten divisions of Edentia life, you would quickly classify the first three as vegetable and the last three as animal, but you would be utterly unable to comprehend the nature of the intervening four groups of prolific and fascinating forms of life.

Even the distinctively animal life is very different from that of the evolutionary worlds, so different that it is quite impossible to portray to mortal minds the unique character and affectionate nature of these nonspeaking creatures. There are thousands upon thousands of living creatures which your imagination could not possibly picture. The whole animal creation is of an entirely different order from the gross animal species of the evolutionary planets. But all this animal life is most intelligent and exquisitely serviceable, and all the various species are surprisingly gentle and touchingly companionable. There are no carnivorous creatures on such architectural worlds; there is nothing in all Edentia to make any living being afraid.

The vegetable life is also very different from that of Urantia, consisting of both material and morontia varieties. The material growths have a characteristic green coloration, but the morontia equivalents of vegetative life have a violet or orchid tinge of varying hue and reflection. Such morontia vegetation is purely an energy growth; when eaten there is no residual portion.

Being endowed with ten divisions of physical life, not to mention the morontia variations, these architectural worlds provide tremendous possibilities for the biologic beautification of the landscape and of the material and the morontia structures. The celestial artisans direct the native spornagia in this extensive work of botanic decoration and biologic embellishment. Whereas your artists must resort to inert paint and lifeless marble to portray their concepts, the celestial artisans and the univitatia more frequently utilize living materials to represent their ideas and to capture their ideals.

If you enjoy the flowers, shrubs, and trees of Urantia, then will you feast your eyes upon the botanical beauty and the floral grandeur of the supernal gardens of Edentia. But it is beyond my powers of description to undertake to convey to the mortal mind an adequate concept of these beauties of the heavenly worlds. Truly, eye has not seen such glories as await your arrival on these worlds of the mortal-ascension adventure.

7. THE UNIVITATIA

Univitatia are the permanent citizens of Edentia and its associated worlds, all seven hundred seventy worlds surrounding the
constellation headquarters being under their supervision. These
children of the Creator Son and the Creative Spirit are projected on a
plane of existence in between the material and the spiritual, but they
are not morontia creatures. The natives of each of the seventy major
spheres of Edentia possess different visible forms, and the morontia
mortals have their morontia forms attuned to correspond with the
ascending scale of the univitatia each time they change residence
from one Edentia sphere to another as they pass successively from
world number one to world number seventy.

Spiritually, the univitatia are alike; intellectually, they vary as
do mortals; in form, they much resemble the morontia state of existence, and they are created to function in seventy diverse orders
of personality. Each of these orders of univitatia exhibits ten major
variations of intellectual activity, and each of these varying intellectual types presides over the special training and cultural schools
of progressive occupational or practical socialization on some one
of the ten satellites which swing around each of the major Edentia
worlds.

These seven hundred minor worlds are technical spheres of
practical education in the working of the entire local universe and
are open to all classes of intelligent beings. These training schools of
special skill and technical knowledge are not conducted exclusively
for ascending mortals, although morontia students constitute by far
the largest group of all those who attend these courses of training.
When you are received on any one of the seventy major worlds of
social culture, you are immediately given clearance for each of the
ten surrounding satellites.

In the various courtesy colonies, ascending morontia mortals
predominate among the reversion directors, but the univitatia represent the largest group associated with the Nebadon corps of celestial
artisans. In all Orvonton no extra-Havona beings excepting the
Uversa abandonters can equal the univitatia in artistic skill, social
adaptability, and co-ordinating cleverness.

These citizens of the constellation are not actually members of
the artisan corps, but they freely work with all groups and contribute

much to making the constellation worlds the chief spheres for the realization of the magnificent artistic possibilities of transition culture. They do not function beyond the confines of the constellation headquarters worlds.

8. THE EDENTIA TRAINING WORLDS

The physical endowment of Edentia and its surrounding spheres is well-nigh perfect; they could hardly equal the spiritual grandeur of the spheres of Salvington, but they far surpass the glories of the training worlds of Jerusem. All these Edentia spheres are energized directly by the universal space currents, and their enormous power systems, both material and morontial, are expertly supervised and distributed by the constellation centers, assisted by a competent corps of Master Physical Controllers and Morontia Power Supervisors.

The time spent on the seventy training worlds of transition morontia culture associated with the Edentia age of mortal ascension, is the most settled period in an ascending mortal's career up to the status of a finaliter; this is really the typical morontia life. While you are re-keyed each time you pass from one major cultural world to another, you retain the same morontia body, and there are no periods of personality unconsciousness.

Your sojourn on Edentia and its associated spheres will be chiefly occupied with the mastery of group ethics, the secret of pleasant and profitable interrelationship between the various universe and superuniverse orders of intelligent personalities.

On the mansion worlds you completed the unification of the evolving mortal personality; on the system capital you attained Jerusem citizenship and achieved the willingness to submit the self to the disciplines of group activities and coordinated undertakings; but now on the constellation training worlds you are to achieve the real socialization of your evolving morontia personality. This supernal cultural acquirement consists in learning how to:

1. Live happily and work effectively with ten diverse fellow morontians, while ten such groups are associated in companies of one hundred and then federated in corps of one thousand.

2. Abide joyfully and co-operate heartily with ten univitatia, who, though similar intellectually to morontia beings, are very dif-

ferent in every other way. And then must you function with this group of ten as it co-ordinates with ten other families, which are in turn confederated into a corps of one thousand univitatia.

3. Achieve simultaneous adjustment to both fellow morontians and these host univitatia. Acquire the ability voluntarily and effectively to co-operate with your own order of beings in close working association with a somewhat dissimilar group of intelligent creatures.

4. While thus socially functioning with beings like and unlike yourself, achieve intellectual harmony with, and make vocational adjustment to, both groups of associates.

5. While attaining satisfactory socialization of the personality on intellectual and vocational levels, further perfect the ability to live in intimate contact with similar and slightly dissimilar beings with ever-lessening irritability and ever-diminishing resentment. The reversion directors contribute much to this latter attainment through their group-play activities.

6. Adjust all of these various socialization techniques to the furtherance of the progressive co-ordination of the Paradise-ascension career; augment universe insight by enhancing the ability to grasp the eternal goal-meanings concealed within these seemingly insignificant time-space activities.

7. And then, climax all of these procedures of multi-socialization with the concurrent enhancement of spiritual insight as it pertains to the augmentation of all phases of personal endowment through group spiritual association and morontia co-ordination. Intellectually, socially, and spiritually two moral creatures do not merely double their personal potentials of universe achievement by partnership technique; they more nearly quadruple their attainment and accomplishment possibilities.

We have portrayed Edentia socialization as an association of a morontia mortal with a univitatia family group consisting of ten intellectually dissimilar individuals concomitant with a similar association with ten fellow morontians. But on the first seven major worlds only one ascending mortal lives with ten univitatia. On the second group of seven major worlds two mortals abide with each native group of ten, and so on up until, on the last group of seven major spheres, ten morontia beings are domiciled with ten univitatia. As you learn how better to socialize with the univitatia, you will practice such improved ethics in your relations with your fellow morontia progressors.

As ascending mortals you will enjoy your sojourn on the progress worlds of Edentia, but you will not experience that personal thrill of satisfaction which characterizes your initial contact with universe affairs on the system headquarters or your farewell touch with these realities on the final worlds of the universe capital.

9. Citizenship on Edentia

After graduation from world number seventy, ascending mortals take up residence on Edentia. Ascenders now, for the first time, attend the "assemblies of Paradise" and hear the story of their far-flung career as it is depicted by the Faithful of Days, the first of the Supreme Trinity-origin Personalities they have met.

This entire sojourn on the constellation training worlds, culminating in Edentia citizenship, is a period of true and heavenly bliss for the morontia progressors. Throughout your sojourn on the system worlds you were evolving from a near-animal to a morontia creature; you were more material than spiritual. On the Salvington spheres you will be evolving from a morontia being to the status of a true spirit; you will be more spiritual than material. But on Edentia, ascenders are midway between their former and their future estates, midway in their passage from evolutionary animal to ascending spirit. During your whole stay on Edentia and its worlds you are "as the angels"; you are constantly progressing but all the while maintaining a general and a typical morontia status.

This constellation sojourn of an ascending mortal is the most uniform and stabilized epoch in the entire career of morontia progression. This experience constitutes the prespirit socialization training of the ascenders. It is analogous to the prefinaliter spiritual experience of Havona and to the preabsonite training on Paradise.

Ascending mortals on Edentia are chiefly occupied with the assignments on the seventy progressive univitatia worlds. They also serve in varied capacities on Edentia itself, mainly in conjunction with the constellation program concerned with group, racial, national, and planetary welfare. The Most Highs are not so much engaged in fostering individual advancement on the inhabited worlds; they rule in the kingdoms of men rather than in the hearts of individuals.

And on that day when you are prepared to leave Edentia for the Salvington career, you will pause and look back on one of the most beautiful and most refreshing of all your epochs of training this side of Paradise. But the glory of it all augments as you ascend inward and achieve increased capacity for enlarged appreciation of divine meanings and spiritual values.

[Sponsored by Malavatia Melchizedek.]

ELEVEN

LOCAL SYSTEM HEADQUARTERS

(PAPER 46)

JERUSEM, the headquarters of Satania, is an average capital of a local system, and aside from numerous irregularities occasioned by the Lucifer rebellion and the bestowal of Michael on Urantia, it is typical of similar spheres. Your local system has passed through some stormy experiences, but it is at present being administered most efficiently, and as the ages pass, the results of disharmony are being slowly but surely eradicated. Order and good will are being restored, and the conditions on Jerusem are more and more approaching the heavenly status of your traditions, for the system headquarters is truly the heaven visualized by the majority of twentieth-century religious believers.

1. PHYSICAL ASPECTS OF JERUSEM

Jerusem is divided into one thousand latitudinal sectors and ten thousand longitudinal zones. The sphere has seven major capitals and seventy minor administrative centers. The seven sectional capitals are concerned with diverse activities, and the System Sovereign is present in each at least once a year.

The standard mile of Jerusem is equivalent to about seven Urantia miles. The standard weight, the "gradant," is built up through the decimal system from the mature ultimaton and represents almost exactly ten ounces of your weight. The Satania day equals three days of Urantia time, less one hour, four minutes, and fifteen seconds, that being the time of the axial revolution of Jerusem. The system year consists of one hundred Jerusem days. The time of the system is broadcast by the master chronoldeks.

The energy of Jerusem is superbly controlled and circulates about the sphere in the zone channels, which are directly fed from the energy charges of space and expertly administered by the Master Physical Controllers. The natural resistance to the passage of these energies through the physical channels of conduction yields the heat required for the production of the equable temperature of Jerusem. The full-light temperature is maintained at about 70 degrees Fahrenheit, while during the period of light recession it falls to a little lower than 50 degrees.

The lighting system of Jerusem should not be so difficult for you to comprehend. There are no days and nights, no seasons of heat and cold. The power transformers maintain one hundred thousand centers from which rarefied energies are projected upward through the planetary atmosphere, undergoing certain changes, until they reach the electric air-ceiling of the sphere; and then these energies are reflected back and down as a gentle, sifting, and even light of about the intensity of Urantia sunlight when the sun is shining overhead at ten o'clock in the morning.

Under such conditions of lighting, the light rays do not seem to come from one place; they just sift out of the sky, emanating equally from all space directions. This light is very similar to natural sunlight except that it contains very much less heat. Thus it will be recognized that such headquarters worlds are not luminous in space; if Jerusem were very near Urantia, it would not be visible.

The gases which reflect this light-energy from the Jerusem upper ionosphere back to the ground are very similar to those in the Urantia upper air belts which are concerned with the auroral phenomena of your so-called northern lights, although these are produced by different causes. On Urantia it is this same gas shield which prevents the escape of the terrestrial broadcast waves, reflecting them earthward when they strike this gas belt in their direct outward flight. In this way broadcasts are held near the surface as they journey through the air around your world.

This lighting of the sphere is uniformly maintained for seventy-five per cent of the Jerusem day, and then there is a gradual recession until, at the time of minimum illumination, the light is about that of your full moon on a clear night. This is the quiet hour for all Jerusem. Only the broadcast-receiving stations are in operation during this period of rest and rehabilitation.

Jerusem receives faint light from several near-by suns—a sort of brilliant starlight—but it is not dependent on them; worlds like Jerusem are not subject to the vicissitudes of sun disturbances, neither are they confronted with the problem of a cooling or dying sun.

The seven transitional study worlds and their forty-nine satellites are heated, lighted, energized, and watered by the Jerusem technique.

2. PHYSICAL FEATURES OF JERUSEM

On Jerusem you will miss the rugged mountain ranges of Urantia and other evolved worlds since there are neither earthquakes nor rainfalls, but you will enjoy the beauteous highlands and other unique variations of topography and landscape. Enormous areas of Jerusem are preserved in a "natural state," and the grandeur of such districts is quite beyond the powers of human imagination.

There are thousands upon thousands of small lakes but no raging rivers nor expansive oceans. There is no rainfall, neither storms nor blizzards, on any of the architectural worlds, but there is the daily precipitation of the condensation of moisture during the time of lowest temperature attending the light recession. (The dew point is higher on a three-gas world than on a two-gas planet like Urantia.) The physical plant life and the morontia world of living things both require moisture, but this is largely supplied by the subsoil system of circulation which extends all over the sphere, even up to the very tops of the highlands. This water system is not entirely subsurface, for there are many canals interconnecting the sparkling lakes of Jerusem.

The atmosphere of Jerusem is a three-gas mixture. This air is very similar to that of Urantia with the addition of a gas adapted to the respiration of the morontia order of life. This third gas in no way unfits the air for the respiration of animals or plants of the material orders.

The transportation system is allied with the circulatory streams of energy movement, these main energy currents being located at ten-mile intervals. By adjustment of physical mechanisms the material beings of the planet can proceed at a pace varying from two to five hundred miles per hour. The transport birds fly at about one hundred miles an hour. The air mechanisms of the Material Sons travel around five hundred miles per hour. Material and early morontia beings must utilize these mechanical means of transport, but spirit personalities proceed by liaison with the superior forces and spirit sources of energy.

Jerusem and its associated worlds are endowed with the ten standard divisions of physical life characteristic of the architectural spheres of Nebadon. And since there is no organic evolution on Jerusem, there are no conflicting forms of life, no struggle for existence, no survival of the fittest. Rather is there a creative adaptation which foreshadows the beauty, the harmony, and the perfection of the eternal worlds of the central and divine universe. And in all this creative perfection there is the most amazing intermingling of physical and of morontia life, artistically contrasted by the celestial artisans and their fellows.

Jerusem is indeed a foretaste of paradisiacal glory and grandeur. But you can never hope to gain an adequate idea of these glorious architectural worlds by any attempted description. There is so little that can be compared with aught on your world, and even then the things of Jerusem so transcend the things of Urantia that the comparison is almost grotesque. Until you actually arrive on Jerusem, you can hardly entertain anything like a true concept of the heavenly worlds, but that is not so long a time in the future when your coming experience on the system capital is compared with your sometime arrival on the more remote training spheres of the universe, the superuniverse, and of Havona.

The manufacturing or laboratory sector of Jerusem is an extensive domain, one which Urantians would hardly recognize since it has no smoking chimneys; nevertheless, there is an intricate material economy associated with these special worlds, and there is a perfection of mechanical technique and physical achievement which would astonish and even awe your most experienced chemists and inventors. Pause to consider that this first world of detention in the Paradise journey is far more material than spiritual. Throughout your stay on Jerusem and its transition worlds you are far nearer your earth life of material things than your later life of advancing spirit existence.

Mount Seraph is the highest elevation on Jerusem, almost fifteen thousand feet, and is the point of departure for all transport seraphim. Numerous mechanical developments are used in providing initial energy for escaping the planetary gravity and overcoming the air resistance. A seraphic transport departs every three seconds of Urantia time throughout the light period and, sometimes, far into the recession. The transporters take off at about twenty-five standard miles per second of Urantia time and do not attain standard velocity until they are over two thousand miles away from Jerusem.

Transports arrive on the crystal field, the so-called sea of glass. Around this area are the receiving stations for the various orders of beings who traverse space by seraphic transport. Near the polar crystal receiving station for student visitors you may ascend the pearly observatory and view the immense relief map of the entire headquarters planet.

3. THE JERUSEM BROADCASTS

The superuniverse and Paradise-Havona broadcasts are received on Jerusem in liaison with Salvington and by a technique involving the polar crystal, the sea of glass. In addition to provisions for the reception of these extra-Nebadon communications, there are three distinct groups of receiving stations. These separate but tricircular groups of stations are adjusted to the reception of broadcasts from the local worlds, from the constellation headquarters, and from the capital of the local universe. All these broadcasts are automatically displayed so as to be discernible by all types of beings present in the central broadcast amphitheater; of all preoccupations for an ascendant mortal on Jerusem, none is more engaging and engrossing than that of listening in on the never-ending stream of universe space reports.

This Jerusem broadcast-receiving station is encircled by an enormous amphitheater, constructed of scintillating materials largely unknown on Urantia and seating over five billion beings—material and morontia—besides accommodating innumerable spirit personalities. It is the favorite diversion for all Jerusem to spend their leisure at the broadcast station, there to learn of the welfare and state of the universe. And this is the only planetary activity which is not slowed down during the recession of light.

At this broadcast-receiving amphitheater the Salvington messages are coming in continuously. Near by, the Edentia word of the Most High Constellation Fathers is received at least once a day. Periodically the regular and special broadcasts of Uversa are relayed through Salvington, and when Paradise messages are in reception, the entire population is assembled around the sea of glass, and the Uversa friends add the reflectivity phenomena to the technique of the Paradise broadcast so that everything heard becomes visible. And it is in this manner that continual foretastes of advancing beauty and grandeur are afforded the mortal survivors as they journey inward on the eternal adventure.

The Jerusem sending station is located at the opposite pole of the sphere. All broadcasts to the individual worlds are relayed from the system capitals except the Michael messages, which sometimes go direct to their destinations over the archangels' circuit.

4. Residential and Administrative Areas

Considerable portions of Jerusem are assigned as residential areas, while other portions of the system capital are given over to the necessary administrative functions involving the supervision of the affairs of 619 inhabited spheres, 56 transitional-culture worlds, and the system capital itself. On Jerusem and in Nebadon these arrangements are designed as follows:

1. *The circles*—the nonnative residential areas.
2. *The squares*—the system executive-administrative areas.
3. *The rectangles*—the rendezvous of the lower native life.
4. *The triangles*—the local or Jerusem administrative areas.

This arrangement of the system activities into circles, squares, rectangles, and triangles is common to all the system capitals of Nebadon. In another universe an entirely different arrangement might prevail. These are matters determined by the diverse plans of the Creator Sons.

Our narrative of these residential and administrative areas takes no account of the vast and beautiful estates of the Material Sons of God, the permanent citizens of Jerusem, neither do we mention numerous other fascinating orders of spirit and near-spirit creatures. For example: Jerusem enjoys the efficient services of the spironga of design for system function. These beings are devoted to spiritual ministry in behalf of the supermaterial residents and visitors. They are a wonderful group of intelligent and beautiful beings who are the transition servants of the higher morontia creatures and of the morontia helpers who labor for the upkeep and embellishment of all morontia creations. They are on Jerusem what the midway creatures are on Urantia, midway helpers functioning between the material and the spiritual.

The system capitals are unique in that they are the only worlds which exhibit well-nigh perfectly all three phases of universe existence: the material, the morontial, and the spiritual. Whether you are a material, morontia, or spirit personality, you will feel at home on Jerusem; so also do the combined beings, such as the midway creatures and the Material Sons.

Jerusem has great buildings of both material and morontia types, while the embellishment of the purely spiritual zones is no less exquisite and replete. If I only had words to tell you of the morontia counterparts of the marvelous physical equipment of Jerusem! If I could only go on to portray the sublime grandeur and exquisite perfection of the spiritual appointments of this headquarters world! Your most imaginative concept of perfection of beauty and repleteness of appointment would hardly approach these grandeurs. And Jerusem is but the first step on the way to the supernal perfection of Paradise beauty.

5. The Jerusem Circles

The residential reservations assigned to the major groups of universe life are designated the Jerusem circles. Those circle groups which find mention in these narratives are the following:

1. The circles of the Sons of God.

2. The circles of the angels and higher spirits.

3. The circles of the Universe Aids, including the creature-trinitized sons not assigned to the Trinity Teacher Sons.

4. The circles of the Master Physical Controllers.

5. The circles of the assigned ascending mortals, including the midway creatures.

6. The circles of the courtesy colonies.

7. The circles of the Corps of the Finality.

Each of these residential groupings consists of seven concentric and successively elevated circles. They are all constructed along the same lines but are of different sizes and are fashioned of differing materials. They are all surrounded by far-reaching enclosures, which mount up to form extensive promenades entirely encompassing every group of seven concentric circles.

1. *Circles of the Sons of God.* Though the Sons of God possess a social planet of their own, one of the transitional-culture worlds, they also occupy these extensive domains on Jerusem. On their transitional-culture world the ascending mortals freely mingle with all orders of divine sonship. There you will personally know and love these Sons, but their social life is largely confined to this special world and its satellites. In the Jerusem circles, however, these various groups of sonship may be observed at work. And since morontia vision is of enormous range, you can walk about on the Sons' promenades and overlook the intriguing activities of their numerous orders.

These seven circles of the Sons are concentric and successively elevated so that each of the outer and larger circles overlooks the inner and smaller ones, each being surrounded by a public promenade wall. These walls are constructed of crystal gems of gleaming brightness and are so elevated as to overlook all of their respective residential circles. The many gates—from fifty to one hundred and fifty thousand—which penetrate each of these walls consist of single pearly crystals.

The first circle of the domain of the Sons is occupied by the Magisterial Sons and their personal staffs. Here center all of the plans and immediate activities of the bestowal and adjudicational services of these juridical Sons. It is also through this center that the Avonals of the system maintain contact with the universe.

The second circle is occupied by the Trinity Teacher Sons. In this sacred domain the Daynals and their associates carry forward the training of the newly arrived primary Teacher Sons. And in all of this work they are ably assisted by a division of certain co-ordinates of the Brilliant Evening Stars. The creature-trinitized sons occupy a sector of the Daynal circle. The Trinity Teacher Sons come the nearest to being the personal representatives of the Universal Father in a local system; they are at least Trinity-origin beings. This second circle is a domain of extraordinary interest to all the peoples of Jerusem.

The third circle is devoted to the Melchizedeks. Here the system chiefs reside and supervise the almost endless activities of these versatile Sons. From the first of the mansion worlds on through all the Jerusem career of ascending mortals, the Melchizedeks are foster fathers and ever-present advisers. It would not be amiss to say that they are the dominant influence on Jerusem aside from the ever-present activities of the Material Sons and Daughters.

The fourth circle is the home of the Vorondadeks and all other orders of the visiting and observer Sons who are not otherwise provided for. The Most High Constellation Fathers take up their abode in this circle when on visits of inspection to the local system. Perfectors of Wisdom, Divine Counselors, and Universal Censors all reside in this circle when on duty in the system.

The fifth circle is the abode of the Lanonandeks, the sonship order of the System Sovereigns and the Planetary Princes. The three groups mingle as one when at home in this domain. The system reserves are held in this circle, while the System Sovereign has a temple situated at the center of the governing group of structures on administration hill.

The sixth circle is the tarrying place of the system Life Carriers. All orders of these Sons are here assembled, and from here they go forth on their world assignments.

The seventh circle is the rendezvous of the ascending sons, those assigned mortals who may be temporarily functioning on the system headquarters, together with their seraphic consorts. All ex-mortals above the status of Jerusem citizens and below that of finaliters are reckoned as belonging to the group having its headquarters in this circle.

These circular reservations of the Sons occupy an enormous area, and until nineteen hundred years ago there existed a great open space at its center. This central region is now occupied by the Michael memorial, completed some five hundred years ago. Four hundred and ninety-five years ago, when this temple was dedicated, Michael was present in person, and all Jerusem heard the touching story of the Master Son's bestowal on Urantia, the least of Satania. The Michael memorial is now the center of all activities embraced in the modified management of the system occasioned by Michael's bestowal, including most of the more recently transplanted Salvington activities. The memorial staff consists of over one million personalities.

2. *The circles of the angels.* Like the residential area of the Sons, these circles of the angels consist of seven concentric and successively elevated circles, each overlooking the inner areas.

The first circle of the angels is occupied by the Higher Personalities of the Infinite Spirit who may be stationed on the headquarters world—Solitary Messengers and their associates. The second circle is dedicated to the messenger hosts, Technical Advisers, companions, inspectors, and recorders as they may chance to function on Jerusem from time to time. The third circle is held by the ministering spirits of the higher orders and groupings.

The fourth circle is held by the administrator seraphim, and the seraphim serving in a local system like Satania are an "innumerable host of angels." The fifth circle is occupied by the planetary seraphim, while the sixth is the home of the transition ministers. The seventh circle is the tarrying sphere of certain unrevealed orders of seraphim. The recorders of all these groups of angels do not sojourn with their fellows, being domiciled in the Jerusem temple of records. All records are preserved in triplicate in this threefold hall of archives. On a system headquarters, records are always preserved in material, in morontia, and in spirit form.

These seven circles are surrounded by the exhibit panorama of Jerusem, five thousand standard miles in circumference, which is devoted to the presentation of the advancing status of the peopled worlds of Satania and is constantly revised so as to truly represent up-to-date conditions on the individual planets. I doubt not that this vast promenade overlooking the circles of the angels will be the first sight of Jerusem to claim your attention when you are permitted extended leisure on your earlier visits.

These exhibits are in charge of the native life of Jerusem, but they are assisted by the ascenders from the various Satania worlds who are tarrying on Jerusem en route to Edentia. The portrayal of planetary conditions and world progress is effected by many methods, some known to you, but mostly by techniques unknown on Urantia. These exhibits occupy the outer edge of this vast wall. The remainder of the promenade is almost entirely open, being highly and magnificently embellished.

3. *The circles of the Universe Aids* have the headquarters of the Evening Stars situated in the enormous central space. Here is located the system headquarters of Galantia, the associate head of this powerful group of superangels, being the first commissioned of all the ascendant Evening Stars. This is one of the most magnificent of all the administrative sectors of Jerusem, even though it is among the more recent constructions. This center is fifty miles in diameter. The Galantia headquarters is a monolithic cast crystal, wholly transparent. These material-morontia crystals are greatly appreciated by both morontia and material beings. The created Evening Stars exert their influence all over Jerusem, being possessed of such extrapersonality attributes. The entire world has been rendered spiritually fragrant since so many of their activities were transferred here from Salvington.

4. *The circles of the Master Physical Controllers.* The various orders of the Master Physical Controllers are concentrically arranged around the vast temple of power, wherein presides the power chief of the system in association with the chief of the Morontia Power Supervisors. This temple of power is one of two sectors on Jerusem where ascending mortals and midway creatures are not permitted. The other one is the dematerializing sector in the area of the Material Sons, a series of laboratories wherein the transport seraphim transform material beings into a state quite like that of the morontia order of existence.

5. *The circles of the ascending mortals.* The central area of the circles of the ascending mortals is occupied by a group of 619 planetary memorials representative of the inhabited worlds of the system, and these structures periodically undergo extensive changes. It is the privilege of the mortals from each world to agree, from time to

time, upon certain of the alterations or additions to their planetary memorials. Many changes are even now being made in the Urantia structures. The center of these 619 temples is occupied by a working model of Edentia and its many worlds of ascendant culture. This model is forty miles in diameter and is an actual reproduction of the Edentia system, true to the original in every detail.

Ascenders enjoy their Jerusem services and take pleasure in observing the techniques of other groups. Everything done in these various circles is open to the full observation of all Jerusem.

The activities of such a world are of three distinct varieties: work, progress, and play. Stated otherwise, they are: service, study, and relaxation. The composite activities consist of social intercourse, group entertainment, and divine worship. There is great educational value in mingling with diverse groups of personalities, orders very different from one's own fellows.

6. *The circles of the courtesy colonies.* The seven circles of the courtesy colonies are graced by three enormous structures: the vast astronomic observatory of Jerusem, the gigantic art gallery of Satania, and the immense assembly hall of the reversion directors, the theater of morontia activities devoted to rest and recreation.

The celestial artisans direct the spornagia and provide the host of creative decorations and monumental memorials which abound in every place of public assembly. The studios of these artisans are among the largest and most beautiful of all the matchless structures of this wonderful world. The other courtesy colonies maintain extensive and beautiful headquarters. Many of these buildings are constructed wholly of crystal gems. All the architectural worlds abound in crystals and the so-called precious metals.

7. *The circles of the finaliters* have a unique structure at the center. And this same vacant temple is found on every system headquarters world throughout Nebadon. This edifice on Jerusem is sealed with the insignia of Michael, and it bears this inscription: "Undedicated to the seventh stage of spirit—to the eternal assignment." Gabriel placed the seal on this temple of mystery, and none but Michael can or may break the seal of sovereignty affixed by the Bright and Morning Star. Some day you shall look upon this silent temple, even though you may not penetrate its mystery.

Other Jerusem circles: In addition to these residential circles there are on Jerusem numerous additional designated abodes.

6. The Executive-Administrative Squares

The executive-administrative divisions of the system are located in the immense departmental squares, one thousand in number. Each administrative unit is divided into one hundred subdivisions of ten subgroups each. These one thousand squares are clustered in ten grand divisions, thus constituting the following ten administrative departments:

1. Physical maintenance and material improvement, the domains of physical power and energy.

2. Arbitration, ethics, and administrative adjudication.

3. Planetary and local affairs.

4. Constellation and universe affairs.

5. Education and other Melchizedek activities.

6. Planetary and system physical progress, the scientific domains of Satania activities.

7. Morontia affairs.

8. Pure spirit activities and ethics.

9. Ascendant ministry.

10. Grand universe philosophy.

These structures are transparent; hence all system activities can be viewed even by student visitors.

7. The Rectangles—the Spornagia

The one thousand *rectangles* of Jerusem are occupied by the lower native life of the headquarters planet, and at their center is situated the vast circular headquarters of the spornagia.

On Jerusem you will be amazed by the agricultural achievements of the wonderful spornagia. There the land is cultivated largely for aesthetic and ornamental effects. The spornagia are the landscape gardeners of the headquarters worlds, and they are both original and artistic in their treatment of the open spaces of Jerusem. They utilize both animals and numerous mechanical contrivances in the culture of the soil. They are intelligently expert in the employment of the power agencies of their realms as well as in the utilization of numerous orders of their lesser brethren of the lower animal creations, many of which are provided them on these special worlds. This order of animal life is now largely directed by the ascending midway creatures from the evolutionary spheres.

Spornagia are not Adjuster indwelt. They do not possess survival souls, but they do enjoy long lives, sometimes to the extent of forty to fifty thousand standard years. Their number is legion, and they afford physical ministry to all orders of universe personalities requiring material service.

Although spornagia neither possess nor evolve survival souls, though they do not have personality, nevertheless, they do evolve an individuality which can experience reincarnation. When, with the passing of time, the physical bodies of these unique creatures deteriorate from usage and age, their creators, in collaboration with the Life Carriers, fabricate new bodies in which the old spornagia re-establish their residences.

Spornagia are the only creatures in all the universe of Nebadon who experience this or any other sort of reincarnation. They are only reactive to the first five of the adjutant mind-spirits; they are not responsive to the spirits of worship and wisdom. But the five-adjutant mind equivalates to a totality or sixth reality level, and it is this factor which persists as an experiential identity.

I am quite without comparisons in undertaking to describe these useful and unusual creatures as there are no animals on the evolutionary worlds comparable to them. They are not evolutionary beings, having been projected by the Life Carriers in their present form and status. They are bisexual and procreate as they are required to meet the needs of a growing population.

Perhaps I can best suggest to Urantia minds something of the nature of these beautiful and serviceable creatures by saying that they embrace the combined traits of a faithful horse and an affectionate dog and manifest an intelligence exceeding that of the highest type of chimpanzee. And they are very beautiful, as judged by the physical standards of Urantia. They are most appreciative of the attentions shown them by the material and semimaterial sojourners on these architectural worlds. They have a vision which permits them to recognize—in addition to material beings—the morontia creations, the lower angelic orders, midway creatures, and some of the lower orders of spirit personalities. They do not comprehend worship of the Infinite, nor do they grasp the import of the Eternal, but they do, through affection for their masters, join in the outward spiritual devotions of their realms.

There are those who believe that, in a future universe age, these faithful spornagia will escape from their animal level of existence and attain a worthy evolutionary destiny of progressive intellectual growth and even spiritual achievement.

8. THE JERUSEM TRIANGLES

The purely local and routine affairs of Jerusem are directed from the one hundred *triangles*. These units are clustered around the ten marvelous structures domiciling the local administration of Jerusem. The triangles are surrounded by the panoramic depiction of the system headquarters history. At present there is an erasure of over two standard miles in this circular story. This sector will be restored upon the readmission of Satania into the constellation family. Every provision for this event has been made by the decrees of Michael, but the tribunal of the Ancients of Days has not yet finished the adjudication of the affairs of the Lucifer rebellion. Satania may not come back into the full fellowship of Norlatiadek so long as it harbors archrebels, high created beings who have fallen from light into darkness.

When Satania can return to the constellation fold, then will come up for consideration the readmission of the isolated worlds into the system family of inhabited planets, accompanied by their restoration to the spiritual communion of the realms. But even if Urantia were restored to the system circuits, you would still be embarrassed by the fact that your whole system rests under a Norlatiadek quarantine partially segregating it from all other systems.

But ere long, the adjudication of Lucifer and his associates will restore the Satania system to the Norlatiadek constellation, and subsequently, Urantia and the other isolated spheres will be restored to the Satania circuits, and again will such worlds enjoy the privileges of interplanetary communication and intersystem communion.

There will come an end for rebels and rebellion. The Supreme Rulers are merciful and patient, but the law of deliberately nourished evil is universally and unerringly executed. "The wages of sin is death"—eternal obliteration.

[Presented by an Archangel of Nebadon.]

TWELVE

THE SEVEN
MANSION WORLDS

(PAPER 47)

THE Creator Son, when on Urantia, spoke of the "many mansions in the Father's universe." In a certain sense, all fifty-six of the encircling worlds of Jerusem are devoted to the transitional culture of ascending mortals, but the seven satellites of world number one are more specifically known as the mansion worlds.

Transition world number one itself is quite exclusively devoted to ascendant activities, being the headquarters of the finaliter corps assigned to Satania. This world now serves as the headquarters for more than one hundred thousand companies of finaliters, and there are one thousand glorified beings in each of these groups.

When a system is settled in light and life, and as the mansion worlds one by one cease to serve as mortal-training stations, they are taken over by the increasing finaliter population which accumulates in these older and more highly perfected systems.

The seven mansion worlds are in charge of the morontia supervisors and the Melchizedeks. There is an acting governor on each world who is directly responsible to the Jerusem rulers. The Uversa conciliators maintain headquarters on each of the mansion worlds, while adjoining is the local rendezvous of the Technical Advisers. The reversion directors and celestial artisans maintain group headquarters on each of these worlds. The spironga function from mansion world number two onward, while all seven, in common with the other transitional-culture planets and the headquarters world, are abundantly provided with spornagia of standard creation.

1. The Finaliter's World

Although only finaliters and certain groups of salvaged children and their caretakers are resident on transitional world number one, provision is made for the entertainment of all classes of spirit beings, transition mortals, and student visitors. The spornagia, who function on all of these worlds, are hospitable hosts to all beings whom they can recognize. They have a vague feeling concerning the finaliters but cannot visualize them. They must regard them much as you do the angels in your present physical state.

Though the finaliter world is a sphere of exquisite physical beauty and extraordinary morontia embellishment, the great spirit abode located at the center of activities, the temple of the finaliters, is not visible to the unaided material or early morontia vision. But the energy transformers are able to visualize many of these realities to ascending mortals, and from time to time they do thus function, as on the occasions of the class assemblies of the mansion world students on this cultural sphere.

All through the mansion world experience you are in a way spiritually aware of the presence of your glorified brethren of Paradise attainment, but it is very refreshing, now and then, actually to perceive them as they function in their headquarters abodes. You will not spontaneously visualize finaliters until you acquire true spirit vision.

On the first mansion world all survivors must pass the requirements of the parental commission from their native planets. The present Urantia commission consists of twelve parental couples, recently arrived, who have had mortal experience in rearing three or more children to the pubescent age. Service on this commission is rotational and is for only ten years as a rule. All who fail to satisfy these commissioners as to their parental experience must further qualify by service in the homes of the Material Sons on Jerusem or in part in the probationary nursery on the finaliters' world.

But irrespective of parental experience, mansion world parents who have growing children in the probation nursery are given every opportunity to collaborate with the morontia custodians of such children regarding their instruction and training. These parents are permitted to journey there for visits as often as four times a year. And it is one of the most touchingly beautiful scenes of all the ascending career to observe the mansion world parents embrace their material offspring on the occasions of their periodic pilgrimages to the finaliter world. While one or both parents may leave a mansion world ahead of the child, they are quite often contemporary for a season.

No ascending mortal can escape the experience of rearing children—their own or others—either on the material worlds or subsequently on the finaliter world or on Jerusem. Fathers must pass through this essential experience just as certainly as mothers. It is an unfortunate and mistaken notion of modern peoples on Urantia that child culture is largely the task of mothers. Children need fathers as well as mothers, and fathers need this parental experience as much as do mothers.

2. The Probationary Nursery

The infant-receiving schools of Satania are situated on the finaliter world, the first of the Jerusem transition-culture spheres. These infant-receiving schools are enterprises devoted to the nurture and training of the children of time, including those who have died on the evolutionary worlds of space before the acquirement of individual status on the universe records. In the event of the survival of either or both of such a child's parents, the guardian of destiny deputizes her associated cherubim as the custodian of the child's potential identity, charging the cherubim with the responsibility of delivering this undeveloped soul into the hands of the Mansion World Teachers in the probationary nurseries of the morontia worlds.

It is these same deserted cherubim who, as Mansion World Teachers, under the supervision of the Melchizedeks, maintain such extensive educational facilities for the training of the probationary wards of the finaliters. These wards of the finaliters, these infants of ascending mortals, are always personalized as of their exact physical status at the time of death except for reproductive potential. This awakening occurs at the exact time of the parental arrival on the first mansion world. And then are these children given every opportunity, as they are, to choose the heavenly way just as they would have made such a choice on the worlds where death so untimely terminated their careers.

On the nursery world, probationary creatures are grouped according to whether or not they have Adjusters, for the Adjusters come to indwell these material children just as on the worlds of time. Children of pre-Adjuster ages are cared for in families of five, ranging in ages from one year and under up to approximately five years, or that age when the Adjuster arrives.

All children on the evolving worlds who have Thought Adjusters, but who before death had not made a choice concerning the Paradise career, are also repersonalized on the finaliter world of the system, where they likewise grow up in the families of the Material Sons and their associates as do those little ones who arrived without Adjusters, but who will subsequently receive the Mystery Monitors after attaining the requisite age of moral choice.

The Adjuster-indwelt children and youths on the finaliter world are also reared in families of five, ranging in ages from six to fourteen; approximately, these families consist of children whose ages are six, eight, ten, twelve, and fourteen. Any time after sixteen, if final choice has been made, they translate to the first mansion world and begin their Paradise ascent. Some make a choice before this age and go on to the ascension spheres, but very few children under sixteen years of age, as reckoned by Urantia standards, will be found on the mansion worlds.

The guardian seraphim attend these youths in the probationary nursery on the finaliter world just as they spiritually minister to mortals on the evolutionary planets, while the faithful spornagia minister to their physical necessities. And so do these children grow up on the transition world until such time as they make their final choice.

When material life has run its course, if no choice has been made for the ascendant life, or if these children of time definitely decide against the Havona adventure, death automatically terminates their probationary careers. There is no adjudication of such cases; there is no resurrection from such a second death. They simply become as though they had not been.

But if they choose the Paradise path of perfection, they are immediately made ready for translation to the first mansion world, where many of them arrive in time to join their parents in the Havona ascent. After passing through Havona and attaining the Deities, these salvaged souls of mortal origin constitute the permanent ascendant citizenship of Paradise. These children who have been deprived of the valuable and essential evolutionary experience on the worlds of mortal nativity are not mustered into the Corps of the Finality.

3. THE FIRST MANSION WORLD

On the mansion worlds the resurrected mortal survivors resume their lives just where they left off when overtaken by death. When you go from Urantia to the first mansion world, you will notice considerable change, but if you had come from a more normal and progressive sphere of time, you would hardly notice the difference except for the fact that you were in possession of a different body; the tabernacle of flesh and blood has been left behind on the world of nativity.

The very center of all activities on the first mansion world is the resurrection hall, the enormous temple of personality assembly. This gigantic structure consists of the central rendezvous of the seraphic destiny guardians, the Thought Adjusters, and the archangels of the resurrection. The Life Carriers also function with these celestial beings in the resurrection of the dead.

The mortal-mind transcripts and the active creature-memory patterns as transformed from the material levels to the spiritual are the individual possession of the detached Thought Adjusters; these spiritized factors of mind, memory, and creature personality are forever a part of such Adjusters. The creature mind-matrix and the passive potentials of identity are present in the morontia soul intrusted to the keeping of the seraphic destiny guardians. And it is the reuniting of the morontia-soul trust of the seraphim and the spirit-mind trust of the Adjuster that reassembles creature personality and constitutes resurrection of a sleeping survivor.

If a transitory personality of mortal origin should never be thus reassembled, the spirit elements of the nonsurviving mortal creature would forever continue as an integral part of the individual experiential endowment of the onetime indwelling Adjuster.

From the Temple of New Life there extend seven radial wings, the resurrection halls of the mortal races. Each of these structures is devoted to the assembly of one of the seven races of time. There are one hundred thousand personal resurrection chambers in each of these seven wings terminating in the circular class assembly halls, which serve as the awakening chambers for as many as one million individuals. These halls are surrounded by the personality assembly chambers of the blended races of the normal post-Adamic worlds. Regardless of the technique which may be employed on the individual worlds of time in connection with special or dispensational resurrections, the real and conscious reassembly of actual and complete personality takes place in the resurrection halls of mansonia number one. Throughout all eternity you will recall the profound memory impressions of your first witnessing of these resurrection mornings.

From the resurrection halls you proceed to the Melchizedek sector, where you are assigned permanent residence. Then you enter upon ten days of personal liberty. You are free to explore the immediate vicinity of your new home and to familiarize yourself with the program which lies immediately ahead. You also have time to gratify your desire to consult the registry and call upon your loved ones and other earth friends who may have preceded you to these worlds. At the end of your ten-day period of leisure you begin the second step in the Paradise journey, for the mansion worlds are actual training spheres, not merely detention planets.

On mansion world number one (or another in case of advanced status) you will resume your intellectual training and spiritual development at the exact level whereon they were interrupted by death. Between the time of planetary death or translation and resurrection on the mansion world, mortal man gains absolutely nothing aside from experiencing the fact of survival. You begin over there right where you leave off down here.

Almost the entire experience of mansion world number one pertains to deficiency ministry. Survivors arriving on this first of the detention spheres present so many and such varied defects of creature character and deficiencies of mortal experience that the major activities of the realm are occupied with the correction and cure of these manifold legacies of the life in the flesh on the material evolutionary worlds of time and space.

The sojourn on mansion world number one is designed to develop mortal survivors at least up to the status of the post-Adamic dispensation on the normal evolutionary worlds. Spiritually, of course, the mansion world students are far in advance of such a state of mere human development.

If you are not to be detained on mansion world number one, at the end of ten days you will enter the translation sleep and proceed to world number two, and every ten days thereafter you will thus advance until you arrive on the world of your assignment.

The center of the seven major circles of the first mansion world administration is occupied by the temple of the Morontia Companions, the personal guides assigned to ascending mortals. These companions are the offspring of the local universe Mother Spirit, and there are several million of them on the morontia worlds of Satania. Aside from those assigned as group companions, you will have much to do with the interpreters and translators, the building custodians, and the excursion supervisors. And all of these companions are most cooperative with those who have to do with developing your personality factors of mind and spirit within the morontia body.

As you start out on the first mansion world, one Morontia Companion is assigned to each company of one thousand ascending mortals, but you will encounter larger numbers as you progress through the seven mansion spheres. These beautiful and versatile beings are companionable associates and charming guides. They are free to accompany individuals or selected groups to any of the transition-culture spheres, including their satellite worlds. They are the excursion guides and leisure associates of all ascending mortals. They often accompany survivor groups on periodic visits to Jerusem, and on any day you are there, you can go to the registry sector of the system capital and meet ascending mortals from all seven of the mansion worlds since they freely journey back and forth between their residential abodes and the system headquarters.

4. THE SECOND MANSION WORLD

It is on this sphere that you are more fully inducted into the mansonia life. The groupings of the morontia life begin to take form; working groups and social organizations start to function, communities take on formal proportions, and the advancing mortals inaugurate new social orders and governmental arrangements.

Spirit-fused survivors occupy the mansion worlds in common with the Adjuster-fused ascending mortals. While the various orders of celestial life differ, they are all friendly and fraternal. In all the worlds of ascension you will find nothing comparable to human intolerance and the discriminations of inconsiderate caste systems.

As you ascend the mansion worlds one by one, they become more crowded with the morontia activities of advancing survivors. As you go forward, you will recognize more and more of the Jerusem features added to the mansion worlds. The sea of glass makes its appearance on the second mansonia.

A newly developed and suitably adjusted morontia body is acquired at the time of each advance from one mansion world to another. You go to sleep with the seraphic transport and awake with the new but undeveloped body in the resurrection halls, much as when you first arrived on mansion world number one except that the Thought Adjuster does not leave you during these transit sleeps between the mansion worlds. Your personality remains intact after you once pass from the evolutionary worlds to the initial mansion world.

Your Adjuster memory remains fully intact as you ascend the morontia life. Those mental associations that were purely animalistic and wholly material naturally perished with the physical brain, but everything in your mental life which was worth while, and which had survival value, was counterparted by the Adjuster and is retained as a part of personal memory all the way through the ascendant career. You will be conscious of all your worth-while experiences as you advance from one mansion world to another and from one section of the universe to another—even to Paradise.

Though you have morontia bodies, you continue, through all seven of these worlds, to eat, drink, and rest. You partake of the morontia order of food, a kingdom of living energy unknown on the material worlds. Both food and water are fully utilized in the morontia body; there is no residual waste. Pause to consider: Mansonia number one is a very material sphere, presenting the early beginnings of the morontia regime. You are still a near human and not far removed from the limited viewpoints of mortal life, but each world discloses definite progress. From sphere to sphere you grow less material, more intellectual, and slightly more spiritual. The spiritual progress is greatest on the last three of these seven progressive worlds.

Biological deficiencies were largely made up on the first mansion world. There defects in planetary experiences pertaining to sex life, family association, and parental function were either corrected or were projected for future rectification among the Material Son families on Jerusem.

Mansonia number two more specifically provides for the removal of all phases of intellectual conflict and for the cure of all varieties of mental disharmony. The effort to master the significance of morontia mota, begun on the first mansion world, is here more earnestly continued. The development on mansonia number two compares with the intellectual status of the post-Magisterial Son culture of the ideal evolutionary worlds.

5. The Third Mansion World

Mansonia the third is the headquarters of the Mansion World Teachers. Though they function on all seven of the mansion spheres, they maintain their group headquarters at the center of the school circles of world number three. There are millions of these instructors on the mansion and higher morontia worlds. These advanced and glorified cherubim serve as morontia teachers all the way up from the mansion worlds to the last sphere of local universe ascendant training. They will be among the last to bid you an affectionate adieu when the farewell time draws near, the time when you bid good-bye—at least for a few ages—to the universe of your origin, when you enseraphim for transit to the receiving worlds of the minor sector of the superuniverse.

When sojourning on the first mansion world, you have permission to visit the first of the transition worlds, the headquarters of the finaliters and the system probationary nursery for the nurture of undeveloped evolutionary children. When you arrive on mansonia number two, you receive permission periodically to visit transition world number two, where are located the morontia supervisor headquarters for all Satania and the training schools for the various morontia orders. When you reach mansion world number three, you are immediately granted a permit to visit the third transition sphere, the headquarters of the angelic orders and the home of their various system training schools. Visits to Jerusem from this world are increasingly profitable and are of ever-heightening interest to the advancing mortals.

Mansonia the third is a world of great personal and social achievement for all who have not made the equivalent of these circles of culture prior to release from the flesh on the mortal nativity worlds. On this sphere more positive educational work is begun. The training of the first two mansion worlds is mostly of a deficiency nature—negative—in that it has to do with supplementing the experience of the life in the flesh. On this third mansion world the survivors really begin their progressive morontia culture. The chief purpose of this training is to enhance the understanding of the correlation of morontia mota and mortal logic, the co-ordination of morontia mota and human philosophy. Surviving mortals now gain practical insight into true metaphysics. This is the real introduction to the intelligent comprehension of cosmic meanings and universe interrelationships. The culture of the third mansion world partakes of the nature of the postbestowal Son age of a normal inhabited planet.

6. THE FOURTH MANSION WORLD

When you arrive on the fourth mansion world, you have well entered upon the morontia career; you have progressed a long way from the initial material existence. Now are you given permission to make visits to transition world number four, there to become familiar with the headquarters and training schools of the superangels, including the Brilliant Evening Stars. Through the good offices of these superangels of the fourth transition world the morontia visitors are enabled to draw very close to the various orders of the Sons of God during the periodic visits to Jerusem, for new sectors of the system capital are gradually opening up to the advancing mortals as they make these repeated visits to the headquarters world. New grandeurs are progressively unfolding to the expanding minds of these ascenders.

On the fourth mansonia the individual ascender more fittingly finds his place in the group working and class functions of the morontia life. Ascenders here develop increased appreciation of the broadcasts and other phases of local universe culture and progress.

It is during the period of training on world number four that the ascending mortals are really first introduced to the demands and delights of the true social life of morontia creatures. And it is indeed a new experience for evolutionary creatures to participate in social activities which are predicated neither on personal aggrandizement nor on self-seeking conquest. A new social order is being introduced, one based on the understanding sympathy of mutual appreciation, the unselfish love of mutual service, and the overmastering motivation of the realization of a common and supreme destiny—the Paradise goal of worshipful and divine perfection. Ascenders are all becoming self-conscious of God-knowing, God-revealing, God-seeking, and God-finding.

The intellectual and social culture of this fourth mansion world is comparable to the mental and social life of the post-Teacher Son age on the planets of normal evolution. The spiritual status is much in advance of such a mortal dispensation.

7. THE FIFTH MANSION WORLD

Transport to the fifth mansion world represents a tremendous forward step in the life of a morontia progressor. The experience on this world is a real foretaste of Jerusem life. Here you begin to realize the high destiny of the loyal evolutionary worlds since they may normally progress to this stage during their natural planetary development. The culture of this mansion world corresponds in general to that of the early era of light and life on the planets of normal evolutionary progress. And from this you can understand why it is so arranged that the highly cultured and progressive types of beings who sometimes inhabit these advanced evolutionary worlds are exempt from passing through one or more, or even all, of the mansion spheres.

Having mastered the local universe language before leaving the fourth mansion world, you now devote more time to the perfection of the tongue of Uversa to the end that you may be proficient in both languages before arriving on Jerusem with residential status. All ascending mortals are bilingual from the system headquarters up to Havona. And then it is only necessary to enlarge the superuniverse vocabulary, still additional enlargement being required for residence on Paradise.

Upon arrival on mansonia number five the pilgrim is given permission to visit the transition world of corresponding number, the Sons' headquarters. Here the ascendant mortal becomes personally familiar with the various groups of divine sonship. He has heard of these superb beings and has already met them on Jerusem, but now he comes really to know them.

On the fifth mansonia you begin to learn of the constellation study worlds. Here you meet the first of the instructors who begin to prepare you for the subsequent constellation sojourn. More of this preparation continues on worlds six and seven, while the finishing touches are supplied in the sector of the ascending mortals on Jerusem.

A real birth of cosmic consciousness takes place on mansonia number five. You are becoming universe minded. This is indeed a time of expanding horizons. It is beginning to dawn upon the enlarging minds of the ascending mortals that some stupendous and magnificent, some supernal and divine, destiny awaits all who complete the progressive Paradise ascension, which has been so laboriously but so joyfully and auspiciously begun. At about this point the average mortal ascender begins to manifest bona fide experiential enthusiasm for the Havona ascent. Study is becoming voluntary, unselfish service natural, and worship spontaneous. A real morontia character is budding; a real morontia creature is evolving.

8. THE SIXTH MANSION WORLD

Sojourners on this sphere are permitted to visit transition world number six, where they learn more about the high spirits of the superuniverse, although they are not able to visualize many of these celestial beings. Here they also receive their first lessons in the prospective spirit career which so immediately follows graduation from the morontia training of the local universe.

The assistant System Sovereign makes frequent visits to this world, and the initial instruction is here begun in the technique of universe administration. The first lessons embracing the affairs of a whole universe are now imparted.

This is a brilliant age for ascending mortals and usually witnesses the perfect fusion of the human mind and the divine Adjuster. In potential, this fusion may have occurred previously, but the actual working identity many times is not achieved until the time of the sojourn on the fifth mansion world or even the sixth.

The union of the evolving immortal soul with the eternal and divine Adjuster is signalized by the seraphic summoning of the supervising superangel for resurrected survivors and of the archangel of record for those going to judgment on the third day; and then, in the presence of such a survivor's morontia associates, these messengers of confirmation speak: "This is a beloved son in whom I am well pleased." This simple ceremony marks the entrance of an ascending mortal upon the eternal career of Paradise service.

Immediately upon the confirmation of Adjuster fusion the new morontia being is introduced to his fellows for the first time by his new name and is granted the forty days of spiritual retirement from all routine activities wherein to commune with himself and to choose some one of the optional routes to Havona and to select from the differential techniques of Paradise attainment.

But still are these brilliant beings more or less material; they are far from being true spirits; they are more like supermortals, spiritually speaking, still a little lower than the angels. But they are truly becoming marvelous creatures.

During the sojourn on world number six the mansion world students achieve a status which is comparable with the exalted development characterizing those evolutionary worlds which have normally progressed beyond the initial stage of light and life. The organization of society on this mansonia is of a high order. The shadow of the mortal nature grows less and less as these worlds are ascended one by one. You are becoming more and more adorable as you leave behind the coarse vestiges of planetary animal origin. "Coming up through great tribulation" serves to make glorified mortals very kind and understanding, very sympathetic and tolerant.

9. The Seventh Mansion World

The experience on this sphere is the crowning achievement of the immediate postmortal career. During your sojourn here you will receive the instruction of many teachers, all of whom will co-operate in the task of preparing you for residence on Jerusem. Any discernible differences between those mortals hailing from the isolated and retarded worlds and those survivors from the more advanced and enlightened spheres are virtually obliterated during the sojourn on the seventh mansion world. Here you will be purged of all the remnants of unfortunate heredity, unwholesome environment, and unspiritual planetary tendencies. The last remnants of the "mark of the beast" are here eradicated.

While sojourning on mansonia number seven, permission is granted to visit transition world number seven, the world of the Universal Father. Here you begin a new and more spiritual worship of the unseen Father, a habit you will increasingly pursue all the way up through your long ascending career. You find the Father's temple on this world of transitional culture, but you do not see the Father.

Now begins the formation of classes for graduation to Jerusem. You have gone from world to world as individuals, but now you prepare to depart for Jerusem in groups, although, within certain limits, an ascender may elect to tarry on the seventh mansion world for the purpose of enabling a tardy member of his earthly or mansonia working group to catch up with him.

The personnel of the seventh mansonia assemble on the sea of glass to witness your departure for Jerusem with residential status. Hundreds or thousands of times you may have visited Jerusem, but always as a guest; never before have you proceeded toward the system capital in the company of a group of your fellows who were bidding an eternal farewell to the whole mansonia career as ascending mortals. You will soon be welcomed on the receiving field of the headquarters world as Jerusem citizens.

You will greatly enjoy your progress through the seven dematerializing worlds; they are really demortalizing spheres. You are mostly human on the first mansion world, just a mortal being minus a material body, a human mind housed in a morontia form—a material body of the morontia world but not a mortal house of flesh and blood. You really pass from the mortal state to the immortal status at the time of Adjuster fusion, and by the time you have finished the Jerusem career, you will be full-fledged morontians.

10. JERUSEM CITIZENSHIP

The reception of a new class of mansion world graduates is the signal for all Jerusem to assemble as a committee of welcome. Even the spornagia enjoy the arrival of these triumphant ascenders of evolutionary origin, those who have run the planetary race and finished the mansion world progression. Only the physical controllers and Morontia Power Supervisors are absent from these occasions of rejoicing.

John the Revelator saw a vision of the arrival of a class of advancing mortals from the seventh mansion world to their first heaven, the glories of Jerusem. He recorded: "And I saw as it were a sea of glass mingled with fire; and those who had gained the victory over the beast that was originally in them and over the image that persisted through the mansion worlds and finally over the last mark and trace, standing on the sea of glass, having the harps of God, and singing the song of deliverance from mortal fear and death." (Perfected space communication is to be had on all these worlds; and your anywhere reception of such communications is made possible by carrying the "harp of God," a morontia contrivance compensating for the inability to directly adjust the immature morontia sensory mechanism to the reception of space communications.)

Paul also had a view of the ascendant-citizen corps of perfecting mortals on Jerusem, for he wrote: "But you have come to Mount Zion and to the city of the living God, the heavenly Jerusalem, and to an innumerable company of angels, to the grand assembly of Michael, and to the spirits of just men being made perfect."

After mortals have attained residence on the system headquarters, no more literal resurrections will be experienced. The morontia form granted you on departure from the mansion world career is such as will see you through to the end of the local universe experience. Changes will be made from time to time, but you will retain this same form until you bid it farewell when you emerge as first-stage spirits preparatory for transit to the superuniverse worlds of ascending culture and spirit training.

Seven times do those mortals who pass through the entire mansonia career experience the adjustment sleep and the resurrection awakening. But the last resurrection hall, the final awakening chamber, was left behind on the seventh mansion world. No more will a form-change necessitate the lapse of consciousness or a break in the continuity of personal memory.

The mortal personality initiated on the evolutionary worlds and tabernacled in the flesh—indwelt by the Mystery Monitors and invested by the Spirit of Truth—is not fully mobilized, realized, and unified until that day when such a Jerusem citizen is given clearance for Edentia and proclaimed a true member of the morontia corps of Nebadon—an immortal survivor of Adjuster association, a Paradise ascender, a personality of morontia status, and a true child of the Most Highs.

Mortal death is a technique of escape from the material life in the flesh; and the mansonia experience of progressive life through seven worlds of corrective training and cultural education represents the introduction of mortal survivors to the morontia career, the transition life which intervenes between the evolutionary material existence and the higher spirit attainment of the ascenders of time who are destined to achieve the portals of eternity.

[Sponsored by a Brilliant Evening Star.]

THIRTEEN

THE MORONTIA LIFE

(PAPER 48)

THE Gods cannot—at least they do not—transform a creature of gross animal nature into a perfected spirit by some mysterious act of creative magic. When the Creators desire to produce perfect beings, they do so by direct and original creation, but they never undertake to convert animal-origin and material creatures into beings of perfection in a single step.

The morontia life, extending as it does over the various stages of the local universe career, is the only possible approach whereby material mortals could attain the threshold of the spirit world. What magic could death, the natural dissolution of the material body, hold that such a simple step should instantly transform the mortal and material mind into an immortal and perfected spirit? Such beliefs are but ignorant superstitions and pleasing fables.

Always this morontia transition intervenes between the mortal estate and the subsequent spirit status of surviving human beings. This intermediate state of universe progress differs markedly in the various local creations, but in intent and purpose they are all quite similar. The arrangement of the mansion and higher morontia worlds in Nebadon is fairly typical of the morontia transition regimes in this part of Orvonton.

1. MORONTIA MATERIALS

The morontia realms are the local universe liaison spheres between the material and spiritual levels of creature existence. This morontia life has been known on Urantia since the early days of the Planetary Prince. From time to time this transition state has been taught to mortals, and the concept, in distorted form, has found a place in present-day religions.

The morontia spheres are the transition phases of mortal ascension through the progression worlds of the local universe. Only the seven worlds surrounding the finaliters' sphere of the local systems are called mansion worlds, but all fifty-six of the system transition abodes, in common with the higher spheres around the constellations and the universe headquarters, are called morontia worlds. These creations partake of the physical beauty and the morontia grandeur of the local universe headquarters spheres.

All of these worlds are architectural spheres, and they have just double the number of elements of the evolved planets. Such made-to-order worlds not only abound in the heavy metals and crystals, having one hundred physical elements, but likewise have exactly one hundred forms of a unique energy organization called *morontia material*. The Master Physical Controllers and the Morontia Power Supervisors are able so to modify the revolutions of the primary units of matter and at the same time so to transform these associations of energy as to create this new substance.

The early morontia life in the local systems is very much like that of your present material world, becoming less physical and more truly morontial on the constellation study worlds. And as you advance to the Salvington spheres, you increasingly attain spiritual levels.

The Morontia Power Supervisors are able to effect a union of material and of spiritual energies, thereby organizing a morontia form of materialization which is receptive to the superimposition of a controlling spirit. When you traverse the morontia life of Nebadon, these same patient and skillful Morontia Power Supervisors will successively provide you with 570 morontia bodies, each one a phase of your progressive transformation. From the time of leaving the material worlds until you are constituted a first-stage spirit on Salvington, you will undergo just 570 separate and ascending morontia changes. Eight of these occur in the system, seventy-one in the constellation, and 491 during the sojourn on the spheres of Salvington.

In the days of the mortal flesh the divine spirit indwells you, almost as a thing apart—in reality an invasion of man by the bestowed spirit of the Universal Father. But in the morontia life the spirit will become a real part of your personality, and as you successively pass through the 570 progressive transformations, you ascend from the material to the spiritual estate of creature life.

Paul learned of the existence of the morontia worlds and of the reality of morontia materials, for he wrote, "They have in heaven a better and more enduring substance." And these morontia materials are real, literal, even as in "the city which has foundations, whose builder and maker is God." And each of these marvelous spheres is "a better country, that is, a heavenly one."

2. Morontia Power Supervisors

These unique beings are exclusively concerned with the supervision of those activities which represent a working combination of spiritual and physical or semimaterial energies. They are exclusively devoted to the ministry of morontia progression. Not that they so much minister to mortals during the transition experience, but they rather make possible the transition environment for the progressing morontia creatures. They are the channels of morontia power which sustain and energize the morontia phases of the transition worlds.

Morontia Power Supervisors are the offspring of a local universe Mother Spirit. They are fairly standard in design though differing slightly in nature in the various local creations. They are created for their specific function and require no training before entering upon their responsibilities.

The creation of the first Morontia Power Supervisors is simultaneous with the arrival of the first mortal survivor on the shores of some one of the first mansion worlds in a local universe. They are created in groups of one thousand, classified as follows:

1. Circuit Regulators   400
2. System Co-ordinators   200
3. Planetary Custodians   100
4. Combined Controllers   100
5. Liaison Stabilizers   100
6. Selective Assorters   50
7. Associate Registrars   50

The power supervisors always serve in their native universe. They are directed exclusively by the joint spirit activity of the Universe Son and the Universe Spirit but are otherwise a wholly self-governing group. They maintain headquarters on each of the first mansion worlds of the local systems, where they work in close association with both the physical controllers and the seraphim but function in a world of their own as regards energy manifestation and spirit application.

They also sometimes work in connection with supermaterial phenomena on the evolutionary worlds as ministers of temporary assignment. But they rarely serve on the inhabited planets; neither do they work on the higher training worlds of the superuniverse, being chiefly devoted to the transition regime of morontia progression in a local universe.

1. *Circuit Regulators.* These are the unique beings who co-ordinate physical and spiritual energy and regulate its flow into the segregated channels of the morontia spheres, and these circuits are exclusively planetary, limited to a single world. The morontia circuits are distinct from, and supplementary to, both physical and spiritual circuits on the transition worlds, and it requires millions of these regulators to energize even a system of mansion worlds like that of Satania.

Circuit regulators initiate those changes in material energies which render them subject to the control and regulation of their associates. These beings are morontia power generators as well as circuit regulators. Much as a dynamo apparently generates electricity out of the atmosphere, so do these living morontia dynamos seem to transform the everywhere energies of space into those materials which the morontia supervisors weave into the bodies and life activities of the ascending mortals.

2. *System Co-ordinators.* Since each morontia world has a separate order of morontia energy, it is exceedingly difficult for humans to visualize these spheres. But on each successive transition sphere, mortals will find the plant life and everything else pertaining to the morontia existence progressively modified to correspond with the advancing spiritization of the ascending survivor. And since the energy system of each world is thus individualized, these co-ordinators operate to harmonize and blend such differing power systems into a working unit for the associated spheres of any particular group.

Ascending mortals gradually progress from the physical to the spiritual as they advance from one morontia world to another; hence the necessity for providing an ascending scale of morontia spheres and an ascending scale of morontia forms.

When mansion world ascenders pass from one sphere to another, they are delivered by the transport seraphim to the receivers of the system co-ordinators on the advanced world. Here in those unique temples at the center of the seventy radiating wings wherein are the chambers of transition similar to the resurrection halls on the initial world of reception for earth-origin mortals, the necessary changes in creature form are skillfully effected by the system co-ordinators. These early morontia-form changes require about seven days of standard time for their accomplishment.

3. *Planetary Custodians.* Each morontia world, from the mansion spheres up to the universe headquarters, is in the custody—as regards morontia affairs—of seventy guardians. They constitute the local planetary council of supreme morontia authority. This council grants material for morontia forms to all ascending creatures who land on the spheres and authorizes those changes in creature form which make it possible for an ascender to proceed to the succeeding sphere. After the mansion worlds have been traversed, you will translate from one phase of morontia life to another without having to surrender consciousness. Unconsciousness attends only the earlier metamorphoses and the later transitions from one universe to another and from Havona to Paradise.

4. *Combined Controllers.* One of these highly mechanical beings is always stationed at the center of each administrative unit of a morontia world. A combined controller is sensitive to, and functional with, physical, spiritual, and morontial energies; and with this being there are always associated two system co-ordinators, four circuit regulators, one planetary custodian, one liaison stabilizer, and either an associate registrar or a selective assorter.

5. *Liaison Stabilizers.* These are the regulators of the morontia energy in association with the physical and spirit forces of the realm. They make possible the conversion of morontia energy into morontia material. The whole morontia organization of existence is dependent on the stabilizers. They slow down the energy revolu-

tions to that point where physicalization can occur. But I have no terms with which I can compare or illustrate the ministry of such beings. It is quite beyond human imagination.

6. *Selective Assorters.* As you progress from one class or phase of a morontia world to another, you must be re-keyed or advance-tuned, and it is the task of the selective assorters to keep you in progressive synchrony with the morontia life.

While the basic morontia forms of life and matter are identical from the first mansion world to the last universe transition sphere, there is a functional progression which gradually extends from the material to the spiritual. Your adaptation to this basically uniform but successively advancing and spiritizing creation is effected by this selective re-keying. Such an adjustment in the mechanism of personality is tantamount to a new creation, notwithstanding that you retain the same morontia form.

You may repeatedly subject yourself to the test of these examiners, and as soon as you register adequate spiritual achievement, they will gladly certify you for advanced standing. These progressive changes result in altered reactions to the morontia environment, such as modifications in food requirements and numerous other personal practices.

The selective assorters are also of great service in the grouping of morontia personalities for purposes of study, teaching, and other projects. They naturally indicate those who will best function in temporary association.

7. *Associate Registrars.* The morontia world has its own recorders, who serve in association with the spirit recorders in the supervision and custody of the records and other data indigenous to the morontia creations. The morontia records are available to all orders of personalities.

All morontia transition realms are accessible alike to material and spirit beings. As morontia progressors you will remain in full contact with the material world and with material personalities, while you will increasingly discern and fraternize with spirit beings; and by the time of departure from the morontia regime, you will have seen all orders of spirits with the exception of a few of the higher types, such as Solitary Messengers.

3. Morontia Companions

These hosts of the mansion and morontia worlds are the off-spring of a local universe Mother Spirit. They are created from age to age in groups of one hundred thousand, and in Nebadon there are at present over seventy billion of these unique beings.

Morontia Companions are trained for service by the Melchize-deks on a special planet near Salvington; they do not pass through the central Melchizedek schools. In service they range from the lowest mansion worlds of the systems to the highest study spheres of Salvington, but they are seldom encountered on the inhabited worlds. They serve under the general supervision of the Sons of God and under the immediate direction of the Melchizedeks.

The Morontia Companions maintain ten thousand headquarters in a local universe—on each of the first mansion worlds of the local systems. They are almost wholly a self-governing order and are, in general, an intelligent and loyal group of beings; but every now and then, in connection with certain unfortunate celestial up-heavals, they have been known to go astray. Thousands of these useful creatures were lost during the times of the Lucifer rebellion in Satania. Your local system now has its full quota of these beings, the loss of the Lucifer rebellion having only recently been made up.

There are two distinct types of Morontia Companions; one type is aggressive, the other retiring, but otherwise they are equal in status. They are not sex creatures, but they manifest a touchingly beautiful affection for one another. And while they are hardly com-panionate in the material (human) sense, they are very close of kin to the human races in the order of creature existence. The midway creatures of the worlds are your nearest of kin; then come the mo-rontia cherubim, and after them the Morontia Companions.

These companions are touchingly affectionate and charm-ingly social beings. They possess distinct personalities, and when you meet them on the mansion worlds, after learning to recognize them as a class, you will soon discern their individuality. Mortals all resemble one another; at the same time each of you possesses a distinct and recognizable personality.

Something of an idea of the nature of the work of these Moron-tia Companions may be derived from the following classification of their activities in a local system:

1. *Pilgrim Guardians* are not assigned to specific duties in their association with the morontia progressors. These companions are responsible for the whole of the morontia career and are therefore the co-ordinators of the work of all other morontia and transition ministers.

2. *Pilgrim Receivers and Free Associators.* These are the social companions of the new arrivals on the mansion worlds. One of them will certainly be on hand to welcome you when you awaken on the initial mansion world from the first transit sleep of time, when you experience the resurrection from the death of the flesh into the morontia life. And from the time you are thus formally welcomed on awakening to that day when you leave the local universe as a first-stage spirit, these Morontia Companions are ever with you.

Companions are not assigned permanently to individuals. An ascending mortal on one of the mansion or higher worlds might have a different companion on each of several successive occasions and again might go for long periods without one. It would all depend on the requirements and also on the supply of companions available.

3. *Hosts to Celestial Visitors.* These gracious creatures are dedicated to the entertainment of the superhuman groups of student visitors and other celestials who may chance to sojourn on the transition worlds. You will have ample opportunity to visit within any realm you have experientially attained. Student visitors are allowed on all inhabited planets, even those in isolation.

4. *Co-ordinators and Liaison Directors.* These companions are dedicated to the facilitation of morontia intercourse and to the prevention of confusion. They are the instructors of social conduct and morontia progress, sponsoring classes and other group activities among the ascending mortals. They maintain extensive areas wherein they assemble their pupils and from time to time make requisition on the celestial artisans and the reversion directors for the embellishment of their programs. As you progress, you will come in intimate contact with these companions, and you will grow exceedingly fond of both groups. It is a matter of chance as to whether you will be associated with an aggressive or a retiring type of companion.

5. *Interpreters and Translators.* During the early mansonia career you will have frequent recourse to the interpreters and the

translators. They know and speak all the tongues of a local universe; they are the linguists of the realms.

You will not acquire new languages automatically; you will learn a language over there much as you do down here, and these brilliant beings will be your language teachers. The first study on the mansion worlds will be the tongue of Satania and then the language of Nebadon. And while you are mastering these new tongues, the Morontia Companions will be your efficient interpreters and patient translators. You will never encounter a visitor on any of these worlds but that some one of the Morontia Companions will be able to officiate as interpreter.

6. *Excursion and Reversion Supervisors.* These companions will accompany you on the longer trips to the headquarters sphere and to the surrounding worlds of transition culture. They plan, conduct, and supervise all such individual and group tours about the system worlds of training and culture.

7. *Area and Building Custodians.* Even the material and morontia structures increase in perfection and grandeur as you advance in the mansonia career. As individuals and as groups you are permitted to make certain changes in the abodes assigned as headquarters for your sojourn on the different mansion worlds. Many of the activities of these spheres take place in the open enclosures of the variously designated circles, squares, and triangles. The majority of the mansion world structures are roofless, being enclosures of magnificent construction and exquisite embellishment. The climatic and other physical conditions prevailing on the architectural worlds make roofs wholly unnecessary.

These custodians of the transition phases of ascendant life are supreme in the management of morontia affairs. They were created for this work, and pending the factualization of the Supreme Being, always will they remain Morontia Companions; never do they perform other duties.

As systems and universes are settled in light and life, the mansion worlds increasingly cease to function as transition spheres of morontia training. More and more the finaliters institute their new training regime, which appears to be designed to translate the cosmic consciousness from the present level of the grand universe to that of the future outer universes. The Morontia Companions are destined to function increasingly in association with the finaliters and in numerous other realms not at present revealed on Urantia.

You can forecast that these beings are probably going to contribute much to your enjoyment of the mansion worlds, whether your sojourn is to be long or short. And you will continue to enjoy them all the way up to Salvington. They are not, technically, essential to any part of your survival experience. You could reach Salvington without them, but you would greatly miss them. They are the personality luxury of your ascending career in the local universe.

4. THE REVERSION DIRECTORS

Joyful mirth and the smile-equivalent are as universal as music. There is a morontial and a spiritual equivalent of mirth and laughter. The ascendant life is about equally divided between work and play—freedom from assignment.

Celestial relaxation and superhuman humor are quite different from their human analogues, but we all actually indulge in a form of both; and they really accomplish for us, in our state, just about what ideal humor is able to do for you on Urantia. The Morontia Companions are skillful play sponsors, and they are most ably supported by the reversion directors.

You would probably best understand the work of the reversion directors if they were likened to the higher types of humorists on Urantia, though that would be an exceedingly crude and somewhat unfortunate way in which to try to convey an idea of the function of these directors of change and relaxation, these ministers of the exalted humor of the morontia and spirit realms.

In discussing spirit humor, first let me tell you what it is *not*. Spirit jest is never tinged with the accentuation of the misfortunes of the weak and erring. Neither is it ever blasphemous of the righteousness and glory of divinity. Our humor embraces three general levels of appreciation:

1. *Reminiscent jests.* Quips growing out of the memories of past episodes in one's experience of combat, struggle, and sometimes fearfulness, and ofttimes foolish and childish anxiety. To us, this phase of humor derives from the deep-seated and abiding ability to draw upon the past for memory material with which pleasantly to flavor and otherwise lighten the heavy loads of the present.

2. *Current humor.* The senselessness of much that so often causes us serious concern, the joy at discovering the unimportance of much of our serious personal anxiety. We are most appreciative of this phase of humor when we are best able to discount the anxieties of the present in favor of the certainties of the future.

3. *Prophetic joy.* It will perhaps be difficult for mortals to envisage this phase of humor, but we do get a peculiar satisfaction out of the assurance "that all things work together for good"—for spirits and morontians as well as for mortals. This aspect of celestial humor grows out of our faith in the loving overcare of our superiors and in the divine stability of our Supreme Directors.

But the reversion directors of the realms are not concerned exclusively with depicting the high humor of the various orders of intelligent beings; they are also occupied with the leadership of diversion, spiritual recreation and morontia entertainment. And in this connection they have the hearty co-operation of the celestial artisans.

The reversion directors themselves are not a created group; they are a recruited corps embracing beings ranging from the Havona natives down through the messenger hosts of space and the ministering spirits of time to the morontia progressors from the evolutionary worlds. All are volunteers, giving themselves to the work of assisting their fellows in the achievement of thought change and mind rest, for such attitudes are most helpful in recuperating depleted energies.

When partially exhausted by the efforts of attainment, and while awaiting the reception of new energy charges, there is agreeable pleasure in living over again the enactments of other days and ages. *The early experiences of the race or the order are restful to reminisce.* And that is exactly why these artists are called reversion directors—they assist in reverting the memory to a former state of development or to a less experienced status of being.

All beings enjoy this sort of reversion except those who are inherent Creators, hence automatic self-rejuvenators, and certain highly specialized types of creatures, such as the power centers and the physical controllers, who are always and eternally thoroughly businesslike in all their reactions. These periodic releases from the tension of functional duty are a regular part of life on all worlds throughout the universe of universes but not on the Isle of Paradise. Beings indigenous to the central abode are incapable of depletion and are not, therefore, subject to re-energizing. And with such beings of eternal Paradise perfection there can be no such reversion to evolutionary experiences.

Most of us have come up through lower stages of existence or through progressive levels of our orders, and it is refreshing and in a measure amusing to look back upon certain episodes of our early experience. There is a restfulness in the contemplation of that which is old to one's order, and which lingers as a memory possession of the mind. The future signifies struggle and advancement; it bespeaks work, effort, and achievement; but the past savors of things already mastered and achieved; contemplation of the past permits of relaxation and such a carefree review as to provoke spirit mirth and a morontia state of mind verging on merriment.

Even mortal humor becomes most hearty when it depicts episodes affecting those just a little beneath one's present developmental state, or when it portrays one's supposed superiors falling victim to the experiences which are commonly associated with supposed inferiors. You of Urantia have allowed much that is at once vulgar and unkind to become confused with your humor, but on the whole, you are to be congratulated on a comparatively keen sense of humor. Some of your races have a rich vein of it and are greatly helped in their earthly careers thereby. Apparently you received much in the way of humor from your Adamic inheritance, much more than was secured of either music or art.

All Satania, during times of play, those times when its inhabitants refreshingly resurrect the memories of a lower stage of existence, is edified by the pleasant humor of a corps of reversion directors from Urantia. The sense of celestial humor we have with us always, even when engaged in the most difficult of assignments. It helps to avoid an overdevelopment of the notion of one's selfimportance. But we do not give rein to it freely, as you might say, "have fun," except when we are in recess from the serious assignments of our respective orders.

When we are tempted to magnify our self-importance, if we stop to contemplate the infinity of the greatness and grandeur of our

Makers, our own self-glorification becomes sublimely ridiculous, even verging on the humorous. One of the functions of humor is to help all of us take ourselves less seriously. *Humor is the divine antidote for exaltation of ego.*

The need for the relaxation and diversion of humor is greatest in those orders of ascendant beings who are subjected to sustained stress in their upward struggles. The two extremes of life have little need for humorous diversions. Primitive men have no capacity therefor, and beings of Paradise perfection have no need thereof. The hosts of Havona are naturally a joyous and exhilarating assemblage of supremely happy personalities. On Paradise the quality of worship obviates the necessity for reversion activities. But among those who start their careers far below the goal of Paradise perfection, there is a large place for the ministry of the reversion directors.

The higher the mortal species, the greater the stress and the greater the capacity for humor as well as the necessity for it. In the spirit world the opposite is true: The higher we ascend, the less the need for the diversions of reversion experiences. But proceeding down the scale of spirit life from Paradise to the seraphic hosts, there is an increasing need for the mission of mirth and the ministry of merriment. Those beings who most need the refreshment of periodic reversion to the intellectual status of previous experiences are the higher types of the human species, the morontians, angels, and the Material Sons, together with all similar types of personality.

Humor should function as an automatic safety valve to prevent the building up of excessive pressures due to the monotony of sustained and serious self-contemplation in association with the intense struggle for developmental progress and noble achievement. Humor also functions to lessen the shock of the unexpected impact of fact or of truth, rigid unyielding fact and flexible everliving truth. The mortal personality, never sure as to which will next be encountered, through humor swiftly grasps—sees the point and achieves insight—the unexpected nature of the situation be it fact or be it truth.

While the humor of Urantia is exceedingly crude and most inartistic, it does serve a valuable purpose both as a health insurance and as a liberator of emotional pressure, thus preventing injurious nervous tension and overserious self-contemplation. Humor and play—relaxation—are never reactions of progressive exertion; always are they the echoes of a backward glance, a reminiscence of the past. Even on Urantia and as you now are, you always find it rejuvenating when for a short time you can suspend the exertions of the newer and higher intellectual efforts and revert to the more simple engagements of your ancestors.

The principles of Urantian play life are philosophically sound and continue to apply on up through your ascending life, through the circuits of Havona to the eternal shores of Paradise. As ascendant beings you are in possession of personal memories of all former and lower existences, and without such identity memories of the past there would be no basis for the humor of the present, either mortal laughter or morontia mirth. It is this recalling of past experiences that provides the basis for present diversion and amusement. And so you will enjoy the celestial equivalents of your earthly humor all the way up through your long morontia, and then increasingly spiritual, careers. And that part of God (the Adjuster) which becomes an eternal part of the personality of an ascendant mortal contributes the overtones of divinity to the joyous expressions, even spiritual laughter, of the ascending creatures of time and space.

5. THE MANSION WORLD TEACHERS

The Mansion World Teachers are a corps of deserted but glorified cherubim and sanobim. When a pilgrim of time advances from a trial world of space to the mansion and associated worlds of morontia training, he is accompanied by his personal or group seraphim, the guardian of destiny. In the worlds of mortal existence the seraphim is ably assisted by cherubim and sanobim; but when her mortal ward is delivered from the bonds of the flesh and starts out on the ascendant career, when the postmaterial or morontia life begins, the attending seraphim has no further need of the ministrations of her former lieutenants, the cherubim and sanobim.

These deserted assistants of the ministering seraphim are often summoned to universe headquarters, where they pass into the intimate embrace of the Universe Mother Spirit and then go forth to the system training spheres as Mansion World Teachers. These teachers often visit the material worlds and function from the lowest mansion worlds on up to the highest of the educational spheres connected with the universe headquarters. Upon their own motion they may return to their former associative work with the ministering seraphim.

There are billions upon billions of these teachers in Satania, and their numbers constantly increase because, in the majority of instances, when a seraphim proceeds inward with an Adjuster-fused mortal, both a cherubim and a sanobim are left behind.

Mansion World Teachers, like most of the other instructors, are commissioned by the Melchizedeks. They are generally supervised by the Morontia Companions, but as individuals and as teachers they are supervised by the acting heads of the schools or spheres wherein they may be functioning as instructors.

These advanced cherubim usually work in pairs as they did when attached to the seraphim. They are by nature very near the morontia type of existence, and they are inherently sympathetic teachers of the ascending mortals and most efficiently conduct the program of the mansion world and morontia educational system.

In the schools of the morontia life these teachers engage in individual, group, class, and mass teaching. On the mansion worlds such schools are organized in three general groups of one hundred divisions each: the schools of thinking, the schools of feeling, and the schools of doing. When you reach the constellation, there are added the schools of ethics, the schools of administration, and the schools of social adjustment. On the universe headquarters worlds you will enter the schools of philosophy, divinity, and pure spirituality.

Those things which you might have learned on earth, but which you failed to learn, must be acquired under the tutelage of these faithful and patient teachers. There are no royal roads, short cuts, or easy paths to Paradise. Irrespective of the individual variations of the route, you master the lessons of one sphere before you proceed to another; at least this is true after you once leave the world of your nativity.

One of the purposes of the morontia career is to effect the permanent eradication from the mortal survivors of such animal vestigial traits as procrastination, equivocation, insincerity, problem avoidance, unfairness, and ease seeking. The mansonia life early teaches the young morontia pupils that postponement is in no sense avoidance. After the life in the flesh, time is no longer available as a technique of dodging situations or of circumventing disagreeable obligations.

Beginning service on the lowest of the tarrying spheres, the Mansion World Teachers advance, with experience, through the educational spheres of the system and the constellation to the training worlds of Salvington. They are subjected to no special discipline either before or after their embrace by the Universe Mother Spirit. They have already been trained for their work while serving as seraphic associates on the worlds native to their pupils of mansion world sojourn. They have had actual experience with these advancing mortals on the inhabited worlds. They are practical and sympathetic teachers, wise and understanding instructors, able and efficient guides. They are entirely familiar with the ascendant plans and thoroughly experienced in the initial phases of the progression career.

Many of the older of these teachers, those who have long served on the worlds of the Salvington circuit, are re-embraced by the Universe Mother Spirit, and from this second embrace these cherubim and sanobim emerge with the status of seraphim.

6. MORONTIA WORLD SERAPHIM— TRANSITION MINISTERS

While all orders of angels, from the planetary helpers to the supreme seraphim, minister on the morontia worlds, the transition ministers are more exclusively assigned to these activities. These angels are of the sixth order of seraphic servers, and their ministry is devoted to facilitating the transit of material and mortal creatures from the temporal life in the flesh on into the early stages of morontia existence on the seven mansion worlds.

You should understand that the morontia life of an ascending mortal is really initiated on the inhabited worlds at the conception of the soul, at that moment when the creature mind of moral status is indwelt by the spirit Adjuster. And from that moment on, the mortal soul has potential capacity for supermortal function, even for recognition on the higher levels of the morontia spheres of the local universe.

You will not, however, be conscious of the ministry of the transition seraphim until you attain the mansion worlds, where they labor untiringly for the advancement of their mortal pupils, being assigned for service in the following seven divisions:

1. *Seraphic Evangels.* The moment you consciousize on the mansion worlds, you are classified as evolving spirits in the records of the system. True, you are not yet spirits in reality, but you are no longer mortal or material beings; you have embarked upon the pre-spirit career and have been duly admitted to the morontia life.

On the mansion worlds the seraphic evangels will help you to choose wisely among the optional routes to Edentia, Salvington, Uversa, and Havona. If there are a number of equally advisable routes, these will be put before you, and you will be permitted to select the one that most appeals to you. These seraphim then make recommendations to the four and twenty advisers on Jerusem concerning that course which would be most advantageous for each ascending soul.

You are not given unrestricted choice as to your future course; but you may choose within the limits of that which the transition ministers and their superiors wisely determine to be most suitable for your future spirit attainment. The spirit world is governed on the principle of respecting your freewill choice provided the course you may choose is not detrimental to you or injurious to your fellows.

These seraphic evangels are dedicated to the proclamation of the gospel of eternal progression, the triumph of perfection attainment. On the mansion worlds they proclaim the great law of the conservation and dominance of goodness: No act of good is ever wholly lost; it may be long thwarted but never wholly annulled, and it is eternally potent in proportion to the divinity of its motivation.

Even on Urantia they counsel the human teachers of truth and righteousness to adhere to the preaching of "the goodness of God, which leads to repentance," to proclaim "the love of God, which casts out all fear." Even so have these truths been declared on your world:

> The Gods are my caretakers; I shall not stray;
> Side by side they lead me in the beautiful paths and glorious
>   refreshing of life everlasting.
> I shall not, in this Divine Presence, want for food nor thirst for
>   water.
> Though I go down into the valley of uncertainty or ascend up
>   into the worlds of doubt,

Though I move in loneliness or with the fellows of my kind,
Though I triumph in the choirs of light or falter in the solitary
    places of the spheres,
Your good spirit shall minister to me, and your glorious angel
    will comfort me.
Though I descend into the depths of darkness and death itself,
I shall not doubt you nor fear you,
For I know that in the fullness of time and the glory of your
    name
You will raise me up to sit with you on the battlements on high.

That is the story whispered in the night season to the shepherd boy. He could not retain it word for word, but to the best of his memory he gave it much as it is recorded today.

These seraphim are also the evangels of the gospel of perfection attainment for the whole system as well as for the individual ascender. Even now in the young system of Satania their teachings and plans encompass provisions for the future ages when the mansion worlds will no longer serve the mortal ascenders as steppingstones to the spheres on high.

2. *Racial Interpreters.* All races of mortal beings are not alike. True, there is a planetary pattern running through the physical, mental, and spiritual natures and tendencies of the various races of a given world; but there are also distinct racial types, and very definite social tendencies characterize the offspring of these different basic types of human beings. On the worlds of time the seraphic racial interpreters further the efforts of the race commissioners to harmonize the varied viewpoints of the races, and they continue to function on the mansion worlds, where these same differences tend to persist in a measure. On a confused planet, such as Urantia, these brilliant beings have hardly had a fair opportunity to function, but they are the skillful sociologists and the wise ethnic advisers of the first heaven.

You should consider the statement about "heaven" and the "heaven of heavens." The heaven conceived by most of your prophets was the first of the mansion worlds of the local system. When the apostle spoke of being "caught up to the third heaven," he referred to that experience in which his Adjuster was detached during sleep and in this unusual state made a projection to the third of the seven mansion worlds. Some of your wise men saw the vision of the greater heaven, "the heaven of heavens," of which the sevenfold mansion world experience was but the first; the second being Jerusem; the third, Edentia and its satellites; the fourth, Salvington and the surrounding educational spheres; the fifth, Uversa; the sixth, Havona; and the seventh, Paradise.

3. *Mind Planners.* These seraphim are devoted to the effective grouping of morontia beings and to organizing their teamwork on the mansion worlds. They are the psychologists of the first heaven. The majority of this particular division of seraphic ministers have had previous experience as guardian angels to the children of time, but their wards, for some reason, failed to personalize on the mansion worlds or else survived by the technique of Spirit fusion.

It is the task of the mind planners to study the nature, experience, and status of the Adjuster souls in transit through the mansion worlds and to facilitate their grouping for assignment and advancement. But these mind planners do not scheme, manipulate, or otherwise take advantage of the ignorance or other limitations of mansion world students. They are wholly fair and eminently just. They respect your newborn morontia will; they regard you as independent volitional beings, and they seek to encourage your speedy development and advancement. Here you are face to face with true friends and understanding counselors, angels who are really able to help you "to see yourself as others see you" and "to know yourself as angels know you."

Even on Urantia, these seraphim teach the everlasting truth: If your own mind does not serve you well, you can exchange it for the mind of Jesus of Nazareth, who always serves you well.

4. *Morontia Counselors.* These ministers receive their name because they are assigned to teach, direct, and counsel the surviving mortals from the worlds of human origin, souls in transit to the higher schools of the system headquarters. They are the teachers of those who seek insight into the experiential unity of divergent life levels, those who are attempting the integration of meanings and the unification of values. This is the function of philosophy in mortal life, of mota on the morontia spheres.

Mota is more than a superior philosophy; it is to philosophy as two eyes are to one; it has a stereoscopic effect on meanings and values. Material man sees the universe, as it were, with but one eye—flat. Mansion world students achieve cosmic perspective—depth—by superimposing the perceptions of the morontia life upon the perceptions of the physical life. And they are enabled to bring these material and morontial viewpoints into true focus largely through the untiring ministry of their seraphic counselors, who so patiently teach the mansion world students and the morontia progressors. Many of the teaching counselors of the supreme order of seraphim began their careers as advisers of the newly liberated souls of the mortals of time.

5. *Technicians.* These are the seraphim who help new ascenders adjust themselves to the new and comparatively strange environment of the morontia spheres. Life on the transition worlds entails real contact with the energies and materials of both the physical and morontia levels and to a certain extent with spiritual realities. Ascenders must acclimatize to every new morontia level, and in all of this they are greatly helped by the seraphic technicians. These seraphim act as liaisons with the Morontia Power Supervisors and with the Master Physical Controllers and function extensively as instructors of the ascending pilgrims concerning the nature of those energies which are utilized on the transition spheres. They serve as emergency space traversers and perform numerous other regular and special duties.

6. *Recorder-Teachers.* These seraphim are the recorders of the borderland transactions of the spiritual and the physical, of the relationships of men and angels, of the morontia transactions of the lower universe realms. They also serve as instructors regarding the efficient and effective techniques of fact recording. There is an artistry in the intelligent assembly and co-ordination of related data, and this art is heightened in collaboration with the celestial artisans, and even the ascending mortals become thus affiliated with the recording seraphim.

The recorders of all the seraphic orders devote a certain amount of time to the education and training of the morontia progressors. These angelic custodians of the facts of time are the ideal instructors of all fact seekers. Before leaving Jerusem, you will become quite familiar with the history of Satania and its 619 inhabited worlds, and much of this story will be imparted by the seraphic recorders.

These angels are all in the chain of recorders extending from the lowest to the highest custodians of the facts of time and the truths of eternity. Some day they will teach you to seek truth as well as fact, to expand your soul as well as your mind. Even now you should learn to water the garden of your heart as well as to seek for the dry sands of knowledge. Forms are valueless when lessons are learned. No chick may be had without the shell, and no shell is of any worth after the chick is hatched. But sometimes error is so great that its rectification by revelation would be fatal to those slowly emerging truths which are essential to its experiential overthrow. When children have their ideals, do not dislodge them; let them grow. And while you are learning to think as men, you should also be learning to pray as children.

Law is life itself and not the rules of its conduct. Evil is a transgression of law, not a violation of the rules of conduct pertaining to life, which *is* the law. Falsehood is not a matter of narration technique but something premeditated as a perversion of truth. The creation of new pictures out of old facts, the restatement of parental life in the lives of offspring—these are the artistic triumphs of truth. The shadow of a hair's turning, premeditated for an untrue purpose, the slightest twisting or perversion of that which is principle—these constitute falseness. But the fetish of factualized truth, fossilized truth, the iron band of so-called unchanging truth, holds one blindly in a closed circle of cold fact. One can be technically right as to fact and everlastingly wrong in the truth.

7. *Ministering Reserves.* A large corps of all orders of the transition seraphim is held on the first mansion world. Next to the destiny guardians, these transition ministers draw the nearest to humans of all orders of seraphim, and many of your leisure moments will be spent with them. Angels take delight in service and, when unassigned, often minister as volunteers. The soul of many an ascending mortal has for the first time been kindled by the divine fire of the will-to-service through personal friendship with the volunteer servers of the seraphic reserves.

From them you will learn to let pressure develop stability and certainty; to be faithful and earnest and, withal, cheerful; to accept challenges without complaint and to face difficulties and uncertainties without fear. They will ask: If you fail, will you rise indomitably to try anew? If you succeed, will you maintain a well-balanced poise—a stabilized and spiritualized attitude—throughout every effort in the long struggle to break the fetters of material inertia, to attain the freedom of spirit existence?

Even as mortals, so have these angels been father to many disappointments, and they will point out that sometimes your most disappointing disappointments have become your greatest blessings. Sometimes the planting of a seed necessitates its death, the death of your fondest hopes, before it can be reborn to bear the fruits of new life and new opportunity. And from them you will learn to suffer less through sorrow and disappointment, first, by making fewer personal plans concerning other personalities, and then, by accepting your lot when you have faithfully performed your duty.

You will learn that you increase your burdens and decrease the likelihood of success by taking yourself too seriously. Nothing can take precedence over the work of your status sphere—this world or the next. Very important is the work of preparation for the next higher sphere, but nothing equals the importance of the work of the world in which you are actually living. But though the *work* is important, the *self* is not. When you feel important, you lose energy to the wear and tear of ego dignity so that there is little energy left to do the work. Self-importance, not work-importance, exhausts immature creatures; it is the self element that exhausts, not the effort to achieve. You can do important work if you do not become self-important; you can do several things as easily as one if you leave yourself out. Variety is restful; monotony is what wears and exhausts. Day after day is alike—just life or the alternative of death.

7. MORONTIA MOTA

The lower planes of morontia mota join directly with the higher levels of human philosophy. On the first mansion world it is the practice to teach the less advanced students by the parallel technique; that is, in one column are presented the more simple concepts of mota meanings, and in the opposite column citation is made of analogous statements of mortal philosophy.

Not long since, while executing an assignment on the first mansion world of Satania, I had occasion to observe this method of teaching; and though I may not undertake to present the mota content of the lesson, I am permitted to record the twenty-eight statements of human philosophy which this morontia instructor was utilizing as illustrative material designed to assist these new mansion world sojourners in their early efforts to grasp the significance and meaning of mota. These illustrations of human philosophy were:

1.   A display of specialized skill does not signify possession of spiritual capacity. Cleverness is not a substitute for true character.

2.   Few persons live up to the faith which they really have. Unreasoned fear is a master intellectual fraud practiced upon the evolving mortal soul.

3.   Inherent capacities cannot be exceeded; a pint can never hold a quart. The spirit concept cannot be mechanically forced into the material memory mold.

4.   Few mortals ever dare to draw anything like the sum of personality credits established by the combined ministries of nature and grace. The majority of impoverished souls are truly rich, but they refuse to believe it.

5.   Difficulties may challenge mediocrity and defeat the fearful, but they only stimulate the true children of the Most Highs.

6.   To enjoy privilege without abuse, to have liberty without license, to possess power and steadfastly refuse to use it for self-aggrandizement—these are the marks of high civilization.

7.   Blind and unforeseen accidents do not occur in the cosmos. Neither do the celestial beings assist the lower being who refuses to act upon his light of truth.

8.   Effort does not always produce joy, but there is no happiness without intelligent effort.

9.   Action achieves strength; moderation eventuates in charm.

10.   Righteousness strikes the harmony chords of truth, and the melody vibrates throughout the cosmos, even to the recognition of the Infinite.

11.   The weak indulge in resolutions, but the strong act. Life is but a day's work—do it well. The act is ours; the consequences God's.

12.   The greatest affliction of the cosmos is never to have been

afflicted. Mortals only learn wisdom by experiencing tribulation.

13.   Stars are best discerned from the lonely isolation of experiential depths, not from the illuminated and ecstatic mountain tops.

14.   Whet the appetites of your associates for truth; give advice only when it is asked for.

15.   Affectation is the ridiculous effort of the ignorant to appear wise, the attempt of the barren soul to appear rich.

16.   You cannot perceive spiritual truth until you feelingly experience it, and many truths are not really felt except in adversity.

17.   Ambition is dangerous until it is fully socialized. You have not truly acquired any virtue until your acts make you worthy of it.

18.   Impatience is a spirit poison; anger is like a stone hurled into a hornet's nest.

19.   Anxiety must be abandoned. The disappointments hardest to bear are those which never come.

20.   Only a poet can discern poetry in the commonplace prose of routine existence.

21.   The high mission of any art is, by its illusions, to foreshadow a higher universe reality, to crystallize the emotions of time into the thought of eternity.

22.   The evolving soul is not made divine by what it does, but by what it strives to do.

23.   Death added nothing to the intellectual possession or to the spiritual endowment, but it did add to the experiential status the consciousness of *survival*.

24.   The destiny of eternity is determined moment by moment by the achievements of the day by day living. The acts of today are the destiny of tomorrow.

25.   Greatness lies not so much in possessing strength as in making a wise and divine use of such strength.

26.   Knowledge is possessed only by sharing; it is safeguarded by wisdom and socialized by love.

27.   Progress demands development of individuality; mediocrity seeks perpetuation in standardization.

28.   The argumentative defense of any proposition is inversely proportional to the truth contained.

Such is the work of the beginners on the first mansion world while the more advanced pupils on the later worlds are mastering the higher levels of cosmic insight and morontia mota.

8. THE MORONTIA PROGRESSSORS

From the time of graduation from the mansion worlds to the attainment of spirit status in the superuniverse career, ascending mortals are denominated morontia progressors. Your passage through this wonderful borderland life will be an unforgettable experience, a charming memory. It is the evolutionary portal to spirit life and the eventual attainment of creature perfection by which ascenders achieve the goal of time—the finding of God on Paradise.

There is a definite and divine purpose in all this morontia and subsequent spirit scheme of mortal progression, this elaborate universe training school for ascending creatures. It is the design of the Creators to afford the creatures of time a graduated opportunity to master the details of the operation and administration of the grand universe, and this long course of training is best carried forward by having the surviving mortal climb up gradually and by actual participation in every step of the ascent.

The mortal-survival plan has a practical and serviceable objective; you are not the recipients of all this divine labor and painstaking training only that you may survive just to enjoy endless bliss and eternal ease. There is a goal of transcendent service concealed beyond the horizon of the present universe age. If the Gods designed merely to take you on one long and eternal joy excursion, they certainly would not so largely turn the whole universe into one vast and intricate practical training school, requisition a substantial part of the celestial creation as teachers and instructors, and then spend ages upon ages piloting you, one by one, through this gigantic universe school of experiential training. The furtherance of the scheme of mortal progression seems to be one of the chief businesses of the present organized universe, and the majority of innumerable orders of created intelligences are either directly or indirectly engaged in advancing some phase of this progressive perfection plan.

In traversing the ascending scale of living existence from mortal man to the Deity embrace, you actually live the very life of every possible phase and stage of perfected creature existence within the limits of the present universe age. From mortal man to Paradise finaliter embraces all that now can be—encompasses everything presently possible to the living orders of intelligent, perfected finite creature beings. If the future destiny of the Paradise finaliters is service in new universes now in the making, it is assured that in this new and future creation there will be no created orders of experiential beings whose lives will be wholly different from those which mortal finaliters have lived on some world as a part of their ascending training, as one of the stages of their agelong progress from animal to angel and from angel to spirit and from spirit to God.

[Presented by an Archangel of Nebadon.]

FOURTEEN

THE INHABITED WORLDS

(PAPER 49)

ALL mortal-inhabited worlds are evolutionary in origin and nature. These spheres are the spawning ground, the evolutionary cradle, of the mortal races of time and space. Each unit of the ascendant life is a veritable training school for the stage of existence just ahead, and this is true of every stage of man's progressive Paradise ascent; just as true of the initial mortal experience on an evolutionary planet as of the final universe headquarters school of the Melchizedeks, a school which is not attended by ascending mortals until just before their translation to the regime of the superuniverse and the attainment of first-stage spirit existence.

All inhabited worlds are basically grouped for celestial administration into the local systems, and each of these local systems is limited to about one thousand evolutionary worlds. This limitation is by the decree of the Ancients of Days, and it pertains to actual evolutionary planets whereon mortals of survival status are living. Neither worlds finally settled in light and life nor planets in the prehuman stage of life development are reckoned in this group.

Satania itself is an unfinished system containing only 619 inhabited worlds. Such planets are numbered serially in accordance with their registration as inhabited worlds, as worlds inhabited by will creatures. Thus was Urantia given the number 606 *of Satania*, meaning the 606th world in this local system on which the long evolutionary life process culminated in the appearance of human beings. There are thirty-six uninhabited planets nearing the life-endowment stage, and several are now being made ready for the Life Carriers. There are nearly two hundred spheres which are evolving so as to be ready for life implantation within the next few million years.

Not all planets are suited to harbor mortal life. Small ones having a high rate of axial revolution are wholly unsuited for life habitats. In several of the physical systems of Satania the planets revolving around the central sun are too large for habitation, their great mass occasioning oppressive gravity. Many of these enormous spheres have satellites, sometimes a half dozen or more, and these moons are often in size very near that of Urantia, so that they are almost ideal for habitation.

The oldest inhabited world of Satania, world number one, is Anova, one of the forty-four satellites revolving around an enormous dark planet but exposed to the differential light of three neighboring suns. Anova is in an advanced stage of progressive civilization.

1. THE PLANETARY LIFE

The universes of time and space are gradual in development; the progression of life—terrestrial or celestial—is neither arbitrary nor magical. Cosmic evolution may not always be understandable (predictable), but it is strictly nonaccidental.

The biologic unit of material life is the protoplasmic cell, the communal association of chemical, electrical, and other basic energies. The chemical formulas differ in each system, and the technique of living cell reproduction is slightly different in each local universe, but the Life Carriers are always the living catalyzers who initiate the primordial reactions of material life; they are the instigators of the energy circuits of living matter.

All the worlds of a local system disclose unmistakable physical kinship; nevertheless, each planet has its own scale of life, no two worlds being exactly alike in plant and animal endowment. These planetary variations in the system life types result from the decisions of the Life Carriers. But these beings are neither capricious nor whimsical; the universes are conducted in accordance with law and order. The laws of Nebadon are the divine mandates of Salvington, and the evolutionary order of life in Satania is in consonance with the evolutionary pattern of Nebadon.

Evolution is the rule of human development, but the process itself varies greatly on different worlds. Life is sometimes initiated in one center, sometimes in three, as it was on Urantia. On the atmospheric worlds it usually has a marine origin, but not always; much depends on the physical status of a planet. The Life Carriers have great latitude in their function of life initiation.

In the development of planetary life the vegetable form always precedes the animal and is quite fully developed before the animal patterns differentiate. All animal types are developed from the basic patterns of the preceding vegetable kingdom of living things; they are not separately organized.

The early stages of life evolution are not altogether in conformity with your present-day views. *Mortal man is not an evolutionary accident.* There is a precise system, a universal law, which determines the unfolding of the planetary life plan on the spheres of space. Time and the production of large numbers of a species are not the controlling influences. Mice reproduce much more rapidly than elephants, yet elephants evolve more rapidly than mice.

The process of planetary evolution is orderly and controlled. The development of higher organisms from lower groupings of life is not accidental. Sometimes evolutionary progress is temporarily delayed by the destruction of certain favorable lines of life plasm carried in a selected species. It often requires ages upon ages to recoup the damage occasioned by the loss of a single superior strain of human heredity. These selected and superior strains of living protoplasm should be jealously and intelligently guarded when once they make their appearance. And on most of the inhabited worlds these superior potentials of life are valued much more highly than on Urantia.

2. PLANETARY PHYSICAL TYPES

There is a standard and basic pattern of vegetable and animal life in each system. But the Life Carriers are oftentimes confronted with the necessity of modifying these basic patterns to conform to the varying physical conditions which confront them on numerous worlds of space. They foster a generalized system type of mortal creature, but there are seven distinct physical types as well as thousands upon thousands of minor variants of these seven outstanding differentiations:

1. Atmospheric types.

2. Elemental types.

3. Gravity types.

4. Temperature types.

5. Electric types.

6. Energizing types.

7. Unnamed types.

The Satania system contains all of these types and numerous intermediate groups, although some are very sparingly represented.

1. *The atmospheric types.* The physical differences of the worlds of mortal habitation are chiefly determined by the nature of the atmosphere; other influences which contribute to the planetary differentiation of life are relatively minor.

The present atmospheric status of Urantia is almost ideal for the support of the breathing type of man, but the human type can be so modified that it can live on both the superatmospheric and the subatmospheric planets. Such modifications also extend to the animal life, which differs greatly on the various inhabited spheres. There is a very great modification of animal orders on both the sub- and the superatmospheric worlds.

Of the atmospheric types in Satania, about two and one-half per cent are subbreathers, about five per cent superbreathers, and over ninety-one per cent are mid-breathers, altogether accounting for ninety-eight and one-half per cent of the Satania worlds.

Beings such as the Urantia races are classified as mid-breathers; you represent the average or typical breathing order of mortal existence. If intelligent creatures should exist on a planet with an atmosphere similar to that of your near neighbor, Venus, they would belong to the superbreather group, while those inhabiting a planet with an atmosphere as thin as that of your outer neighbor, Mars, would be denominated subbreathers.

If mortals should inhabit a planet devoid of air, like your moon, they would belong to the separate order of nonbreathers. This type represents a radical or extreme adjustment to the planetary environment and is separately considered. Nonbreathers account for the remaining one and one-half per cent of Satania worlds.

2. *The elemental types.* These differentiations have to do with the relation of mortals to water, air, and land, and there are four distinct species of intelligent life as they are related to these habitats. The Urantia races are of the land order.

It is quite impossible for you to envisage the environment which prevails during the early ages of some worlds. These unusual conditions make it necessary for the evolving animal life to remain in its marine nursery habitat for longer periods than on those planets which very early provide a hospitable land-and-atmosphere environment. Conversely, on some worlds of the superbreathers, when the planet is not too large, it is sometimes expedient to provide for a mortal type which can readily negotiate atmospheric passage. These air navigators sometimes intervene between the water and land groups, and they always live in a measure upon the ground, eventually evolving into land dwellers. But on some worlds, for ages they continue to fly even after they have become land-type beings.

It is both amazing and amusing to observe the early civilization of a primitive race of human beings taking shape, in one case, in the air and treetops and, in another, midst the shallow waters of sheltered tropic basins, as well as on the bottom, sides, and shores of these marine gardens of the dawn races of such extraordinary spheres. Even on Urantia there was a long age during which primitive man preserved himself and advanced his primitive civilization by living for the most part in the treetops as did his earlier arboreal ancestors. And on Urantia you still have a group of diminutive mammals (the bat family) that are air navigators, and your seals and whales, of marine habitat, are also of the mammalian order.

In Satania, of the elemental types, seven per cent are water, ten per cent air, seventy per cent land, and thirteen per cent combined land-and-air types. But these modifications of early intelligent creatures are neither human fishes nor human birds. They are of the human and prehuman types, neither superfishes nor glorified birds but distinctly mortal.

3. *The gravity types.* By modification of creative design, intelligent beings are so constructed that they can freely function on spheres both smaller and larger than Urantia, thus being, in measure, accommodated to the gravity of those planets which are not of ideal size and density.

The various planetary types of mortals vary in height, the average in Nebadon being a trifle under seven feet. Some of the larger worlds are peopled with beings who are only about two and one-half feet in height. Mortal stature ranges from here on up through the average heights on the average-sized planets to around ten feet on the smaller inhabited spheres. In Satania there is only one race under four feet in height. Twenty per cent of the Satania inhabited worlds are peopled with mortals of the modified gravity types occupying the larger and the smaller planets.

4. *The temperature types.* It is possible to create living beings who can withstand temperatures both much higher and much lower than the life range of the Urantia races. There are five distinct orders of beings as they are classified with reference to heat-regulating mechanisms. In this scale the Urantia races are number three. Thirty per cent of Satania worlds are peopled with races of modified temperature types. Twelve per cent belong to the higher temperature ranges, eighteen per cent to the lower, as compared with Urantians, who function in the mid-temperature group.

5. *The electric types.* The electric, magnetic, and electronic behavior of the worlds varies greatly. There are ten designs of mortal life variously fashioned to withstand the differential energy of the spheres. These ten varieties also react in slightly different ways to the chemical rays of ordinary sunlight. But these slight physical variations in no way affect the intellectual or the spiritual life.

Of the electric groupings of mortal life, almost twenty-three per cent belong to class number four, the Urantia type of existence. These types are distributed as follows: number 1, one per cent; number 2, two per cent; number 3, five per cent; number 4, twenty-three per cent; number 5, twenty-seven per cent; number 6, twenty-four per cent; number 7, eight per cent; number 8, five per cent; number 9, three per cent; number 10, two per cent—in whole percentages.

6. *The energizing types.* Not all worlds are alike in the manner of taking in energy. Not all inhabited worlds have an atmospheric ocean suited to respiratory exchange of gases, such as is present on Urantia. During the earlier and the later stages of many planets, beings of your present order could not exist; and when the respiratory factors of a planet are very high or very low, but when all other

prerequisites to intelligent life are adequate, the Life Carriers often establish on such worlds a modified form of mortal existence, beings who are competent to effect their life-process exchanges directly by means of light-energy and the firsthand power transmutations of the Master Physical Controllers.

There are six differing types of animal and mortal nutrition: The subbreathers employ the first type of nutrition, the marine dwellers the second, the mid-breathers the third, as on Urantia. The superbreathers employ the fourth type of energy intake, while the nonbreathers utilize the fifth order of nutrition and energy. The sixth technique of energizing is limited to the midway creatures.

7. *The unnamed types.* There are numerous additional physical variations in planetary life, but all of these differences are wholly matters of anatomical modification, physiologic differentiation, and electrochemical adjustment. Such distinctions do not concern the intellectual or the spiritual life.

3. Worlds of the Non-Breathers

The majority of inhabited planets are peopled with the breathing type of intelligent beings. But there are also orders of mortals who are able to live on worlds with little or no air. Of the Orvonton inhabited worlds this type amounts to less than seven per cent. In Nebadon this percentage is less than three. In all Satania there are only nine such worlds.

There are so very few of the nonbreather type of inhabited worlds in Satania because this more recently organized section of Norlatiadek still abounds in meteoric space bodies; and worlds without a protective friction atmosphere are subject to incessant bombardment by these wanderers. Even some of the comets consist of meteor swarms, but as a rule they are disrupted smaller bodies of matter.

Millions upon millions of meteorites enter the atmosphere of Urantia daily, coming in at the rate of almost two hundred miles a second. On the nonbreathing worlds the advanced races must do much to protect themselves from meteor damage by making electrical installations which operate to consume or shunt the meteors. Great danger confronts them when they venture beyond these protected zones. These worlds are also subject to disastrous electrical storms of a nature unknown on Urantia. During such times of tremendous energy fluctuation the inhabitants must take refuge in their special structures of protective insulation.

Life on the worlds of the nonbreathers is radically different from what it is on Urantia. The nonbreathers do not eat food or drink water as do the Urantia races. The reactions of the nervous system, the heat-regulating mechanism, and the metabolism of these specialized peoples are radically different from such functions of Urantia mortals. Almost every act of living, aside from reproduction, differs, and even the methods of procreation are somewhat different.

On the nonbreathing worlds the animal species are radically unlike those found on the atmospheric planets. The nonbreathing plan of life varies from the technique of existence on an atmospheric world; even in survival their peoples differ, being candidates for Spirit fusion. Nevertheless, these beings enjoy life and carry forward the activities of the realm with the same relative trials and joys that are experienced by the mortals living on atmospheric worlds. In mind and character the nonbreathers do not differ from other mortal types.

You would be more than interested in the planetary conduct of this type of mortal because such a race of beings inhabits a sphere in close proximity to Urantia.

4. EVOLUTIONARY WILL CREATURES

There are great differences between the mortals of the different worlds, even among those belonging to the same intellectual and physical types, but all mortals of will dignity are erect animals, bipeds.

There are six basic evolutionary races: three primary—red, yellow, and blue; and three secondary—orange, green, and indigo. Most inhabited worlds have all of these races, but many of the three-brained planets harbor only the three primary types. Some local systems also have only these three races.

The average special physical-sense endowment of human beings is twelve, though the special senses of the three-brained mortals are extended slightly beyond those of the one- and two-brained types; they can see and hear considerably more than the Urantia races.

Young are usually born singly, multiple births being the exception, and the family life is fairly uniform on all types of planets. Sex equality prevails on all advanced worlds; male and female are equal in mind endowment and spiritual status. We do not regard a planet as having emerged from barbarism so long as one sex seeks to tyrannize over the other. This feature of creature experience is always greatly improved after the arrival of a Material Son and Daughter.

Seasons and temperature variations occur on all sunlighted and sun-heated planets. Agriculture is universal on all atmospheric worlds; tilling the soil is the one pursuit that is common to the advancing races of all such planets.

Mortals all have the same general struggles with microscopic foes in their early days, such as you now experience on Urantia, though perhaps not so extensive. The length of life varies on the different planets from twenty-five years on the primitive worlds to near five hundred on the more advanced and older spheres.

Human beings are all gregarious, both tribal and racial. These group segregations are inherent in their origin and constitution. Such tendencies can be modified only by advancing civilization and by gradual spiritualization. The social, economic, and governmental problems of the inhabited worlds vary in accordance with the age of the planets and the degree to which they have been influenced by the successive sojourns of the divine Sons.

Mind is the bestowal of the Infinite Spirit and functions quite the same in diverse environments. The mind of mortals is akin, regardless of certain structural and chemical differences which characterize the physical natures of the will creatures of the local systems. Regardless of personal or physical planetary differences, the mental life of all these various orders of mortals is very similar, and their immediate careers after death are very much alike.

But mortal mind without immortal spirit cannot survive. The mind of man is mortal; only the bestowed spirit is immortal. Survival is dependent on spiritualization by the ministry of the Adjuster—on the birth and evolution of the immortal soul; at least, there must not have developed an antagonism towards the Adjuster's mission of effecting the spiritual transformation of the material mind.

5. The Planetary Series of Mortals

It will be somewhat difficult to make an adequate portrayal of the planetary series of mortals because you know so little about them, and because there are so many variations. Mortal creatures may, however, be studied from numerous viewpoints, among which are the following:

1. Adjustment to planetary environment.
2. Brain-type series.
3. Spirit-reception series.
4. Planetary-mortal epochs.
5. Creature-kinship serials.
6. Adjuster-fusion series.
7. Techniques of terrestrial escape.

The inhabited spheres of the seven superuniverses are peopled with mortals who simultaneously classify in some one or more categories of each of these seven generalized classes of evolutionary creature life. But even these general classifications make no provision for such beings as midsoniters nor for certain other forms of intelligent life. The inhabited worlds, as they have been presented in these narratives, are peopled with evolutionary mortal creatures, but there are other life forms.

1. *Adjustment to planetary environment.* There are three general groups of inhabited worlds from the standpoint of the adjustment of creature life to the planetary environment: the normal adjustment group, the radical adjustment group, and the experimental group.

Normal adjustments to planetary conditions follow the general physical patterns previously considered. The worlds of the nonbreathers typify the radical or extreme adjustment, but other types are also included in this group. Experimental worlds are usually ideally adapted to the typical life forms, and on these decimal planets the Life Carriers attempt to produce beneficial variations in the standard life designs. Since your world is an experimental planet, it differs markedly from its sister spheres in Satania; many forms of life have appeared on Urantia that are not found elsewhere; likewise are many common species absent from your planet.

In the universe of Nebadon, all the life-modification worlds are serially linked together and constitute a special domain of universe affairs which is given attention by designated administrators; and all of these experimental worlds are periodically inspected by a corps of universe directors whose chief is the veteran finaliter known in Satania as Tabamantia.

2. *Brain-type series.* The one physical uniformity of mortals is the brain and nervous system; nevertheless, there are three basic organizations of the brain mechanism: the one-, the two-, and the three-brained types. Urantians are of the two-brained type, somewhat more imaginative, adventurous, and philosophical than the one-brained mortals but somewhat less spiritual, ethical, and worshipful than the three-brained orders. These brain differences characterize even the prehuman animal existences.

From the two-hemisphere type of the Urantian cerebral cortex you can, by analogy, grasp something of the one-brained type. The third brain of the three-brained orders is best conceived as an evolvement of your lower or rudimentary form of brain, which is developed to the point where it functions chiefly in control of physical activities, leaving the two superior brains free for higher engagements: one for intellectual functions and the other for the spiritual-counterparting activities of the Thought Adjuster.

While the terrestrial attainments of the one-brained races are slightly limited in comparison with the two-brained orders, the older planets of the three-brained group exhibit civilizations that would astound Urantians, and which would somewhat shame yours by comparison. In mechanical development and material civilization, even in intellectual progress, the two-brained mortal worlds are able to equal the three-brained spheres. But in the higher control of mind and development of intellectual and spiritual reciprocation, you are somewhat inferior.

All such comparative estimates concerning the intellectual progress or the spiritual attainments of any world or group of worlds should in fairness recognize planetary age; much, very much, depends on age, the help of the biologic uplifters, and the subsequent missions of the various orders of the divine Sons.

While the three-brained peoples are capable of a slightly higher planetary evolution than either the one- or two-brained orders, all have the same type of life plasm and carry on planetary activities in very similar ways, much as do human beings on Urantia. These three types of mortals are distributed throughout the worlds of the local systems. In the majority of cases planetary conditions had very little to do with the decisions of the Life Carriers to project these varied orders of mortals on the different worlds; it is a prerogative of the Life Carriers thus to plan and execute.

These three orders stand on an equal footing in the ascension career. Each must traverse the same intellectual scale of development, and each must master the same spiritual tests of progression. The system administration and the constellation overcontrol of these different worlds are uniformly free from discrimination; even the regimes of the Planetary Princes are identical.

3. *Spirit-reception series.* There are three groups of mind design as related to contact with spirit affairs. This classification does not refer to the one-, two-, and three-brained orders of mortals; it refers primarily to gland chemistry, more particularly to the organization of certain glands comparable to the pituitary bodies. The races on some worlds have one gland, on others two, as do Urantians, while on still other spheres the races have three of these unique bodies. The inherent imagination and spiritual receptivity is definitely influenced by this differential chemical endowment.

Of the spirit-reception types, sixty-five per cent are of the second group, like the Urantia races. Twelve per cent are of the first type, naturally less receptive, while twenty-three per cent are more spiritually inclined during terrestrial life. But such distinctions do not survive natural death; all of these racial differences pertain only to the life in the flesh.

4. *Planetary-mortal epochs.* This classification recognizes the succession of temporal dispensations as they affect man's terrestrial status and his reception of celestial ministry.

Life is initiated on the planets by the Life Carriers, who watch over its development until sometime after the evolutionary appearance of mortal man. Before the Life Carriers leave a planet, they duly install a Planetary Prince as ruler of the realm. With this ruler there arrives a full quota of subordinate auxiliaries and ministering helpers, and the first adjudication of the living and the dead is simultaneous with his arrival.

With the emergence of human groupings, this Planetary Prince arrives to inaugurate human civilization and to focalize human society. Your world of confusion is no criterion of the early days of the reign of the Planetary Princes, for it was near the beginning of such an administration on Urantia that your Planetary Prince, Caligastia, cast his lot with the rebellion of the System Sovereign, Lucifer. Your planet has pursued a stormy course ever since.

On a normal evolutionary world, racial progress attains its natural biologic peak during the regime of the Planetary Prince, and shortly thereafter the System Sovereign dispatches a Material Son and Daughter to that planet. These imported beings are of service as biologic uplifters; their default on Urantia further complicated your planetary history.

When the intellectual and ethical progress of a human race has reached the limits of evolutionary development, there comes an Avonal Son of Paradise on a magisterial mission; and later on, when the spiritual status of such a world is nearing its limit of natural attainment, the planet is visited by a Paradise bestowal Son. The chief mission of a bestowal Son is to establish the planetary status, release the Spirit of Truth for planetary function, and thus effect the universal coming of the Thought Adjusters.

Here, again, Urantia deviates: There has never been a magisterial mission on your world, neither was your bestowal Son of the Avonal order; your planet enjoyed the signal honor of becoming the mortal home planet of the Sovereign Son, Michael of Nebadon.

As a result of the ministry of all the successive orders of divine sonship, the inhabited worlds and their advancing races begin to approach the apex of planetary evolution. Such worlds now become ripe for the culminating mission, the arrival of the Trinity Teacher Sons. This epoch of the Teacher Sons is the vestibule to the final planetary age—evolutionary utopia—the age of light and life.

This classification of human beings will receive particular attention in a succeeding paper.

5. *Creature-kinship serials.* Planets are not only organized vertically into systems, constellations, and so on, but the universe administration also provides for horizontal groupings according to type, series, and other relationships. This lateral administration of the universe pertains more particularly to the coordination of activi-

ties of a kindred nature which have been independently fostered on different spheres. These related classes of universe creatures are periodically inspected by certain composite corps of high personalities presided over by long-experienced finaliters.

These kinship factors are manifest on all levels, for kinship serials exist among nonhuman personalities as well as among mortal creatures—even between human and superhuman orders. Intelligent beings are vertically related in twelve great groups of seven major divisions each. The co-ordination of these uniquely related groups of living beings is probably effected by some not fully comprehended technique of the Supreme Being.

6. *Adjuster-fusion series.* The spiritual classification or grouping of all mortals during their prefusion experience is wholly determined by the relation of the personality status to the indwelling Mystery Monitor. Almost ninety per cent of the inhabited worlds of Nebadon are peopled with Adjuster-fusion mortals in contrast with a near-by universe where scarcely more than one half of the worlds harbor beings who are Adjuster-indwelt candidates for eternal fusion.

7. *Techniques of terrestrial escape.* There is fundamentally only one way in which individual human life can be initiated on the inhabited worlds, and that is through creature procreation and natural birth; but there are numerous techniques whereby man escapes his terrestrial status and gains access to the inward moving stream of Paradise ascenders.

6. TERRESTRIAL ESCAPE

All of the differing physical types and planetary series of mortals alike enjoy the ministry of Thought Adjusters, guardian angels, and the various orders of the messenger hosts of the Infinite Spirit. All alike are liberated from the bonds of flesh by the emancipation of natural death, and all alike go thence to the morontia worlds of spiritual evolution and mind progress.

From time to time, on motion of the planetary authorities or the system rulers, special resurrections of the sleeping survivors are conducted. Such resurrections occur at least every millennium of planetary time, when not all but "many of those who sleep in the dust awake." These special resurrections are the occasion for mobilizing special groups of ascenders for specific service in the local universe plan of mortal ascension. There are both practical reasons and sentimental associations connected with these special resurrections.

Throughout the earlier ages of an inhabited world, many are called to the mansion spheres at the special and the millennial resurrections, but most survivors are repersonalized at the inauguration of a new dispensation associated with the advent of a divine Son of planetary service.

1. *Mortals of the dispensational or group order of survival.* With the arrival of the first Adjuster on an inhabited world the guardian seraphim also make their appearance; they are indispensable to terrestrial escape. Throughout the life-lapse period of the sleeping survivors the spiritual values and eternal realities of their newly evolved and immortal souls are held as a sacred trust by the personal or by the group guardian seraphim.

The group guardians of assignment to the sleeping survivors always function with the judgment Sons on their world advents. "He shall send his angels, and they shall gather together his elect from the four winds." With each seraphim of assignment to the repersonalization of a sleeping mortal there functions the returned Adjuster, the same immortal Father fragment that lived in him during the days in the flesh, and thus is identity restored and personality resurrected. During the sleep of their subjects these waiting Adjusters serve on Divinington; they never indwell another mortal mind in this interim.

While the older worlds of mortal existence harbor those highly developed and exquisitely spiritual types of human beings who are virtually exempt from the morontia life, the earlier ages of the animal-origin races are characterized by primitive mortals who are so immature that fusion with their Adjusters is impossible. The reawakening of these mortals is accomplished by the guardian seraphim in conjunction with an individualized portion of the immortal spirit of the Third Source and Center.

Thus are the sleeping survivors of a planetary age repersonalized in the dispensational roll calls. But with regard to the nonsalvable personalities of a realm, no immortal spirit is present to function with the group guardians of destiny, and this constitutes cessation of creature existence. While some of your records have pictured these events as taking place on the planets of mortal death, they all really occur on the mansion worlds.

2. *Mortals of the individual orders of ascension.* The individual progress of human beings is measured by their successive attainment and traversal (mastery) of the seven cosmic circles. These circles of mortal progression are levels of associated intellectual, social, spiritual, and cosmic-insight values. Starting out in the seventh circle, mortals strive for the first, and all who have attained the third immediately have personal guardians of destiny assigned to them. These mortals may be repersonalized in the morontia life independent of dispensational or other adjudications.

Throughout the earlier ages of an evolutionary world, few mortals go to judgment on the third day. But as the ages pass, more and more the personal guardians of destiny are assigned to the advancing mortals, and thus increasing numbers of these evolving creatures are repersonalized on the first mansion world on the third day after natural death. On such occasions the return of the Adjuster signalizes the awakening of the human soul, and this is the repersonalization of the dead just as literally as when the en masse roll is called at the end of a dispensation on the evolutionary worlds.

There are three groups of individual ascenders: The less advanced land on the initial or first mansion world. The more advanced group may take up the morontia career on any of the intermediate mansion worlds in accordance with previous planetary progression. The most advanced of these orders really begin their morontia experience on the seventh mansion world.

3. *Mortals of the probationary-dependent orders of ascension.* The arrival of an Adjuster constitutes identity in the eyes of the universe, and all indwelt beings are on the roll calls of justice. But temporal life on the evolutionary worlds is uncertain, and many die in youth before choosing the Paradise career. Such Adjuster-indwelt children and youths follow the parent of most advanced spiritual status, thus going to the system finaliter world (the probationary nursery) on the third day, at a special resurrection, or at the regular millennial and dispensational roll calls.

Children who die when too young to have Thought Adjusters are repersonalized on the finaliter world of the local systems concomitant with the arrival of either parent on the mansion worlds. A child acquires physical entity at mortal birth, but in the matter of survival all Adjusterless children are reckoned as still attached to their parents.

In due course Thought Adjusters come to indwell these little ones, while the seraphic ministry to both groups of the probationary-dependent orders of survival is in general similar to that of the more advanced parent or is equivalent to that of the parent in case only one survives. Those attaining the third circle, regardless of the status of their parents, are accorded personal guardians.

Similar probation nurseries are maintained on the finaliter spheres of the constellation and the universe headquarters for the Adjusterless children of the primary and secondary modified orders of ascenders.

4. *Mortals of the secondary modified orders of ascension.* These are the progressive human beings of the intermediate evolutionary worlds. As a rule they are not immune to natural death, but they are exempt from passing through the seven mansion worlds.

The less perfected group reawaken on the headquarters of their local system, passing by only the mansion worlds. The intermediate group go to the constellation training worlds; they pass by the entire morontia regime of the local system. Still farther on in the planetary ages of spiritual striving, many survivors awaken on the constellation headquarters and there begin the Paradise ascent.

But before any of these groups may go forward, they must journey back as instructors to the worlds they missed, gaining many experiences as teachers in those realms which they passed by as students. They all subsequently proceed to Paradise by the ordained routes of mortal progression.

5. *Mortals of the primary modified order of ascension.* These mortals belong to the Adjuster-fused type of evolutionary life, but they are most often representative of the final phases of human development on an evolving world. These glorified beings are exempt from passing through the portals of death; they are submitted to Son seizure; they are translated from among the living and appear immediately in the presence of the Sovereign Son on the headquarters of the local universe.

These are the mortals who fuse with their Adjusters during mortal life, and such Adjuster-fused personalities traverse space freely before being clothed with morontia forms. These fused souls go by direct Adjuster transit to the resurrection halls of the higher morontia spheres, where they receive their initial morontia investiture just as do all other mortals arriving from the evolutionary worlds.

This primary modified order of mortal ascension may apply to individuals in any of the planetary series from the lowest to the highest stages of the Adjuster-fusion worlds, but it more frequently functions on the older of these spheres after they have received the benefits of numerous sojourns of the divine Sons.

With the establishment of the planetary era of light and life, many go to the universe morontia worlds by the primary modified order of translation. Further along in the advanced stages of settled existence, when the majority of the mortals leaving a realm are embraced in this class, the planet is regarded as belonging to this series. Natural death becomes decreasingly frequent on these spheres long settled in light and life.

[Presented by a Melchizedek of the Jerusem School of Planetary Administration.]

FIFTEEN

THE SPHERES
OF LIGHT AND LIFE

(PAPER 55)

THE age of light and life is the final evolutionary attainment of a
world of time and space. From the early times of primitive man,
such an inhabited world has passed through the successive planetary
ages—the pre- and the post-Planetary Prince ages, the post-Adamic
age, the post-Magisterial Son age, and the postbestowal Son age.
And then is such a world made ready for the culminating evolutionary attainment, the settled status of light and life, by the ministry of
the successive planetary missions of the Trinity Teacher Sons with
their ever-advancing revelations of divine truth and cosmic wisdom. In these endeavors the Teacher Sons enjoy the assistance of the
Brilliant Evening Stars always, and the Melchizedeks sometimes, in
establishing the final planetary age.

This era of light and life, inaugurated by the Teacher Sons at the
conclusion of their final planetary mission, continues indefinitely
on the inhabited worlds. Each advancing stage of settled status may
be segregated by the judicial actions of the Magisterial Sons into a
succession of dispensations; but all such judicial actions are purely
technical, in no way modifying the course of planetary events.

Only those planets which attain existence in the main circuits
of the superuniverse are assured of continuous survival, but as far as
we know, these worlds settled in light and life are destined to go on

254

throughout the eternal ages of all future time.

There are seven stages in the unfoldment of the era of light and life on an evolutionary world, and in this connection it should be noted that the worlds of the Spirit-fused mortals evolve along lines identical with those of the Adjuster-fusion series. These seven stages of light and life are:

1. The first or planetary stage.

2. The second or system stage.

3. The third or constellation stage.

4. The fourth or local universe stage.

5. The fifth or minor sector stage.

6. The sixth or major sector stage.

7. The seventh or superuniverse stage.

At the conclusion of this narrative these stages of advancing development are described as they relate to the universe organization, but the planetary values of any stage may be attained by any world quite independent of the development of other worlds or of the superplanetary levels of universe administration.

1. THE MORONTIA TEMPLE

The presence of a morontia temple at the capital of an inhabited world is the certificate of the admission of such a sphere to the settled ages of light and life. Before the Teacher Sons leave a world at the conclusion of their terminal mission, they inaugurate this final epoch of evolutionary attainment; they preside on that day when the "holy temple comes down upon earth." This event, signalizing the dawn of the era of light and life, is always honored by the personal presence of the Paradise bestowal Son of that planet, who comes to witness this great day. There in this temple of unparalleled beauty, this bestowal Son of Paradise proclaims the long-time Planetary Prince as the new Planetary Sovereign and invests such a faithful Lanonandek Son with new powers and extended authority over planetary affairs. The System Sovereign is also present and speaks in confirmation of these pronouncements.

A morontia temple has three parts: Centermost is the sanctuary of the Paradise bestowal Son. On the right is the seat of the former Planetary Prince, now Planetary Sovereign; and when present in the temple, this Lanonandek Son is visible to the more spiritual individuals of the realm. On the left is the seat of the acting chief of finaliters attached to the planet.

Although the planetary temples have been spoken of as "coming down from heaven," in reality no actual material is transported from the system headquarters. The architecture of each is worked out in miniature on the system capital, and the Morontia Power Supervisors subsequently bring these approved plans to the planet. Here, in association with the Master Physical Controllers, they proceed to build the morontia temple according to specifications.

The average morontia temple seats about three hundred thousand spectators. These edifices are not used for worship, play, or for receiving broadcasts; they are devoted to the special ceremonies of the planet, such as: communications with the System Sovereign or with the Most Highs, special visualization ceremonies designed to reveal the personality presence of spirit beings, and silent cosmic contemplation. The schools of cosmic philosophy here conduct their graduation exercises, and here also do the mortals of the realm receive planetary recognition for achievements of high social service and for other outstanding attainments.

Such a morontia temple also serves as the place of assembly for witnessing the translation of living mortals to the morontia existence. It is because the translation temple is composed of morontia material that it is not destroyed by the blazing glory of the consuming fire which so completely obliterates the physical bodies of those mortals who therein experience final fusion with their divine Adjusters. On a large world these departure flares are almost continuous, and as the number of translations increases, subsidiary morontia life shrines are provided in different areas of the planet. Not long since I sojourned on a world in the far north whereon twenty-five morontia shrines were functioning.

On presettled worlds, planets without morontia temples, these fusion flashes many times occur in the planetary atmosphere, where the material body of a translation candidate is elevated by the midway creatures and the physical controllers.

2. Death and Translation

Natural, physical death is not a mortal inevitability. The majority of advanced evolutionary beings, citizens on worlds existing in the final era of light and life, do not die; they are translated directly from the life in the flesh to the morontia existence.

This experience of translation from the material life to the morontia state—fusion of the immortal soul with the indwelling Adjuster—increases in frequency commensurate with the evolutionary progress of the planet. At first only a few mortals in each age attain translation levels of spiritual progress, but with the onset of the successive ages of the Teacher Sons, more and more Adjuster fusions occur before the termination of the lengthening lives of these progressing mortals; and by the time of the terminal mission of the Teacher Sons, approximately one quarter of these superb mortals are exempt from natural death.

Farther along in the era of light and life the midway creatures or their associates sense the approaching status of probable soul-Adjuster union and signify this to the destiny guardians, who in turn communicate these matters to the finaliter group under whose jurisdiction this mortal may be functioning; then there is issued the summons of the Planetary Sovereign for such a mortal to resign all planetary duties, bid farewell to the world of his origin, and repair to the inner temple of the Planetary Sovereign, there to await morontia transit, the translation flash, from the material domain of evolution to the morontia level of prespirit progression.

When the family, friends, and working group of such a fusion candidate have forgathered in the morontia temple, they are distributed around the central stage whereon the fusion candidates are resting, meantime freely conversing with their assembled friends. A circle of intervening celestial personalities is arranged to protect the material mortals from the action of the energies manifest at the instant of the "life flash" which delivers the ascension candidate from the bonds of material flesh, thereby doing for such an evolutionary mortal everything that natural death does for those who are thereby delivered from the flesh.

Many fusion candidates may be assembled in the spacious temple at the same time. And what a beautiful occasion when mortals thus forgather to witness the ascension of their loved ones in spiritual flames, and what a contrast to those earlier ages when mortals must commit their dead to the embrace of the terrestrial elements! The scenes of weeping and wailing characteristic of earlier epochs of human evolution are now replaced by ecstatic joy and the sublimest enthusiasm as these God-knowing mortals bid their loved ones a transient farewell as they are removed from their material associations by the spiritual fires of consuming grandeur and ascending glory. On worlds settled in light and life, "funerals" are occasions of supreme joy, profound satisfaction, and inexpressible hope.

The souls of these progressing mortals are increasingly filled with faith, hope, and assurance. The spirit permeating those gathered around the translation shrine resembles that of the joyful friends and relatives who might assemble at a graduating exercise for one of their group, or who might come together to witness the conferring of some great honor upon one of their number. And it would be decidedly helpful if less advanced mortals could only learn to view natural death with something of this same cheerfulness and lightheartedness.

Mortal observers can see nothing of their translated associates subsequent to the fusion flash. Such translated souls proceed by Adjuster transit direct to the resurrection hall of the appropriate morontia-training world. These transactions concerned with the translation of living human beings to the morontia world are supervised by an archangel who was assigned to such a world on the day when it was first settled in light and life.

By the time a world attains the fourth stage of light and life, more than half the mortals leave the planet by translation from among the living. Such diminishment of death continues on and on, but I know of no system whose inhabited worlds, even though long settled in life, are entirely free from natural death as the technique of escape from the bonds of flesh. And until such a high state of planetary evolution is uniformly attained, the morontia-training worlds of the local universe must continue in service as educational and cultural spheres for the evolving morontia progressors. The elimination of death is theoretically possible, but it has not yet occurred according to my observation. Perhaps such a status may be attained during the faraway stretches of the succeeding epochs of the seventh stage of settled planetary life.

The translated souls of the flowering ages of the settled spheres do not pass through the mansion worlds. Neither do they sojourn, as students, on the morontia worlds of the system or constellation. They do not pass through any of the earlier phases of morontia life. They are the only ascending mortals who so nearly escape the morontia transition from material existence to semispirit status. The initial experience of such *Son-seized* mortals in the ascension career is in the services of the progression worlds of the universe headquarters. And from these study worlds of Salvington they go back as teachers to the very worlds they passed by, subsequently going on inward to Paradise by the established route of mortal ascension.

Could you but visit a planet in an advanced stage of development, you would quickly grasp the reasons for providing for the differential reception of ascending mortals on the mansion and higher morontia worlds. You would readily understand that beings passing on from such highly evolved spheres are prepared to resume their Paradise ascent far in advance of the average mortal arriving from a disordered and backward world like Urantia.

No matter from what level of planetary attainment human beings may ascend to the morontia worlds, the seven mansion spheres afford them ample opportunity to gain in experience as teacher-students all of everything which they failed to pass through because of the advanced status of their native planets.

The universe is unfailing in the application of these equalizing techniques designed to insure that no ascender shall be deprived of aught which is essential to his ascension experience.

3. THE GOLDEN AGES

During this age of light and life the world increasingly prospers under the fatherly rule of the Planetary Sovereign. By this time the worlds are progressing under the momentum of one language, one religion, and, on normal spheres, one race. But this age is not perfect. These worlds still have well-appointed hospitals, homes for the care of the sick. There still remain the problems of caring for accidental injuries and the inescapable infirmities attendant upon the decrepitude of old age and the disorders of senility. Disease has not been entirely vanquished, neither have the earth animals been subdued in perfection; but such worlds are like Paradise in comparison with the early times of primitive man during the pre-Planetary

Prince age. You would instinctively describe such a realm—could you be suddenly transported to a planet in this stage of development—as heaven on earth.

Human government in the conduct of material affairs continues to function throughout this age of relative progress and perfection. The public activities of a world in the first stage of light and life which I recently visited were financed by the tithing technique. Every adult worker—and all able-bodied citizens worked at something—paid ten per cent of his income or increase to the public treasury, and it was disbursed as follows:

1. Three per cent was expended in the promotion of truth—science, education, and philosophy.

2. Three per cent was devoted to beauty—play, social leisure, and art.

3. Three per cent was dedicated to goodness—social service, altruism, and religion.

4. One per cent was assigned to the insurance reserves against the risk of incapacity for labor resultant from accident, disease, old age, or unpreventable disasters.

The natural resources of this planet were administered as social possessions, community property.

On this world the highest honor conferred upon a citizen was the order of "supreme service," being the only degree of recognition ever to be granted in the morontia temple. This recognition was bestowed upon those who had long distinguished themselves in some phase of supermaterial discovery or planetary social service.

The majority of social and administrative posts were held jointly by men and women. Most of the teaching was also done jointly; likewise all judicial trusts were discharged by similar associated couples.

On these superb worlds the childbearing period is not greatly prolonged. It is not best for too many years to intervene between the ages of a family of children. When close together in age, children are able to contribute much more to their mutual training. And on these worlds they are magnificently trained by the competitive systems of keen striving in the advanced domains and divisions of diverse achievement in the mastery of truth, beauty, and goodness. Never fear but that even such glorified spheres present plenty of evil, real and potential, which is stimulative of the choosing between truth and error, good and evil, sin and righteousness.

Nevertheless, there is a certain, inevitable penalty attaching to mortal existence on such advanced evolutionary planets. When a settled world progresses beyond the third stage of light and life, all ascenders are destined, before attaining the minor sector, to receive some sort of transient assignment on a planet passing through the earlier stages of evolution.

Each of these successive ages represents advancing achievements in all phases of planetary attainment. In the initial age of light the revelation of truth was enlarged to embrace the workings of the universe of universes, while the Deity study of the second age is the attempt to master the protean concept of the nature, mission, ministry, associations, origin, and destiny of the Creator Sons, the first level of God the Sevenfold.

A planet the size of Urantia, when fairly well settled, would have about one hundred subadministrative centers. These subordinate centers would be presided over by one of the following groups of qualified administrators:

1. Young Material Sons and Daughters brought from the system headquarters to act as assistants to the ruling Adam and Eve.

2. The progeny of the semimortal staff of the Planetary Prince who were procreated on certain worlds for this and other similar responsibilities.

3. The direct planetary progeny of Adam and Eve.

4. Materialized and humanized midway creatures.

5. Mortals of Adjuster-fusion status who, upon their own petition, are temporarily exempted from translation by the order of the Personalized Adjuster of universe chieftainship in order that they may continue on the planet in certain important administrative posts.

6. Specially trained mortals of the planetary schools of administration who have also received the order of supreme service of the morontia temple.

7. Certain elective commissions of three properly qualified citizens who are sometimes chosen by the citizenry by direction of

the Planetary Sovereign in accordance with their special ability to accomplish some definite task which is needful in that particular planetary sector.

The great handicap confronting Urantia in the matter of attaining the high planetary destiny of light and life is embraced in the problems of disease, degeneracy, war, multicolored races, and multilingualism.

No evolutionary world can hope to progress beyond the first stage of settledness in light until it has achieved one language, one religion, and one philosophy. Being of one race greatly facilitates such achievement, but the many peoples of Urantia do not preclude the attainment of higher stages.

4. ADMINISTRATIVE READJUSTMENTS

In the successive stages of settled existence the inhabited worlds make marvelous progress under the wise and sympathetic administration of the volunteer Corps of the Finality, ascenders of Paradise attainment who have come back to minister to their brethren in the flesh. These finaliters are active in co-operation with the Trinity Teacher Sons, but they do not begin their real participation in world affairs until the morontia temple appears on earth.

Upon the formal inauguration of the planetary ministry of the Corps of the Finality, the majority of the celestial hosts withdraw. But the seraphic guardians of destiny continue their personal ministry to the progressing mortals in light; indeed such angels come in ever-increasing numbers throughout the settled ages since larger and larger groups of human beings reach the third cosmic circle of co-ordinate mortal attainment during the planetary life span.

This is merely the first of the successive administrative adjustments which attend the unfolding of the successive ages of increasingly brilliant attainment on the inhabited worlds as they pass from the first to the seventh stage of settled existence.

1. *The first stage of light and life.* A world in this initial settled stage is being administered by three rulers:

a.   The Planetary Sovereign, presently to be advised by a counseling Trinity Teacher Son, in all probability the chief of the terminal corps of such Sons to function on the planet.

b.   The chief of the planetary corps of finaliters.

c. Adam and Eve, who function jointly as the unifiers of the dual leadership of the Prince-Sovereign and the chief of finaliters.

Acting as interpreters for the seraphic guardians and the finaliters are the exalted and liberated midway creatures. One of the last acts of the Trinity Teacher Sons on their terminal mission is to liberate the midwayers of the realm and to promote (or restore) them to advanced planetary status, assigning them to responsible places in the new administration of the settled sphere. Such changes have already been made in the range of human vision as enable mortals to recognize these heretofore invisible cousins of the early Adamic regime. This is made possible by the final discoveries of physical science in liaison with the enlarged planetary functions of the Master Physical Controllers.

The System Sovereign has authority to release midway creatures any time after the first settled stage so that they may humanize in the morontia by the aid of the Life Carriers and the physical controllers and, after receiving Thought Adjusters, start out on their Paradise ascension.

In the third and subsequent stages, some of the midwayers are still functioning, chiefly as contact personalities for the finaliters, but as each stage of light and life is entered, new orders of liaison ministers largely replace the midwayers; very few of them ever remain beyond the fourth stage of light. The seventh stage will witness the coming of the first absonite ministers from Paradise to serve in the places of certain universe creatures.

2. *The second stage of light and life.* This epoch is signalized on the worlds by the arrival of a Life Carrier who becomes the volunteer adviser of the planetary rulers regarding the further efforts to purify and stabilize the mortal race. Thus do the Life Carriers actively participate in the further evolution of the human race—physically, socially, and economically. And then they extend their supervision to the further purification of the mortal stock by the drastic elimination of the retarded and persisting remnants of inferior potential of an intellectual, philosophic, cosmic, and spiritual nature. Those who design and plant life on an inhabited world are fully competent to advise the Material Sons and Daughters, who have full and unquestioned authority to purge the evolving race of all detrimental influences.

From the second stage on throughout the career of a settled planet the Teacher Sons serve as counselors to the finaliters. During such missions they serve as volunteers and not by assignment; and they serve exclusively with the finaliter corps except that, upon the consent of the System Sovereign, they may be had as advisers to the Planetary Adam and Eve.

3. *The third stage of light and life.* During this epoch the inhabited worlds arrive at a new appreciation of the Ancients of Days, the second phase of God the Sevenfold, and the representatives of these superuniverse rulers enter into new relationships with the planetary administration.

In each succeeding age of settled existence the finaliters function in ever-increasing capacities. There exists a close working connection between the finaliters, the Evening Stars (the superangels), and the Trinity Teacher Sons.

During this or the following age a Teacher Son, assisted by the ministering-spirit quartette, becomes attached to the elective mortal chief executive, who now becomes associated with the Planetary Sovereign as joint administrator of world affairs. These mortal chief executives serve for twenty-five years of planetary time, and it is this new development that makes it easy for the Planetary Adam and Eve to secure release from their world of long-time assignment during the following ages.

The ministering-spirit quartettes consist of: the seraphic chief of the sphere, the superuniverse secoraphic counselor, the archangel of translations, and the omniaphim who functions as the personal representative of the Assigned Sentinel stationed on the system headquarters. But these advisers never proffer counsel unless it is asked for.

4. *The fourth stage of light and life.* On the worlds the Trinity Teacher Sons appear in new roles. Assisted by the creature-trinitized sons so long associated with their order, they now come to the worlds as volunteer counselors and advisers to the Planetary Sovereign and his associates. Such couples—Paradise-Havona-trinitized sons and ascender-trinitized sons—represent differing universe viewpoints and diverse personal experiences which are highly serviceable to the planetary rulers.

At any time after this age the Planetary Adam and Eve can petition the Sovereign Creator Son for release from planetary duties in order to begin their Paradise ascent; or they can remain on the planet as directors of the newly appearing order of increasingly spiritual society composed of advanced mortals striving to comprehend the philosophic teachings of the finaliters portrayed by the Brilliant Evening Stars, who are now assigned to these worlds to collaborate in pairs with the seconaphim from the headquarters of the superuniverse.

The finaliters are chiefly engaged in initiating the new and supermaterial activities of society—social, cultural, philosophic, cosmic, and spiritual. As far as we can discern, they will continue this ministry far into the seventh epoch of evolutionary stability, when, possibly, they may go forth to minister in outer space; whereupon we conjecture their places may be taken by absonite beings from Paradise.

5. *The fifth stage of light and life.* The readjustments of this stage of settled existence pertain almost entirely to the physical domains and are of primary concern to the Master Physical Controllers.

6. *The sixth stage of light and life* witnesses the development of new functions of the mind circuits of the realm. Cosmic wisdom seems to become constitutive in the universe ministry of mind.

7. *The seventh stage of light and life.* Early in the seventh epoch the Trinity Teacher counselor of the Planetary Sovereign is joined by a volunteer adviser sent by the Ancients of Days, and later on they will be augmented by a third counselor coming from the superuniverse Supreme Executive.

During this epoch, if not before, Adam and Eve are always relieved of planetary duties. If there is a Material Son in the finaliter corps, he may become associated with the mortal chief executive, and sometimes it is a Melchizedek who volunteers to function in this capacity. If a midwayer is among the finaliters, all of that order remaining on the planet are immediately released.

Upon obtaining release from their agelong assignment, a Planetary Adam and Eve may select careers as follows:

1. They can secure planetary release and from the universe headquarters start out immediately on the Paradise career, receiving Thought Adjusters at the conclusion of the morontia experience.

2. Very often a Planetary Adam and Eve will receive Adjusters while yet serving on a world settled in light concomitant with the receiving of Adjusters by some of their imported pure-line children who have volunteered for a term of planetary service. Subsequently they may all go to universe headquarters and there begin the Paradise career.

3. A Planetary Adam and Eve may elect—as do Material Sons and Daughters from the system capital—to go direct to the midsonite world for a brief sojourn, there to receive their Adjusters.

4. They may decide to return to the system headquarters, there for a time to occupy seats on the supreme court, after which service they will receive Adjusters and begin the Paradise ascent.

5. They may choose to go from their administrative duties back to their native world to serve as teachers for a season and to become Adjuster indwelt at the time of transfer to the universe headquarters.

Throughout all of these epochs the imported assisting Material Sons and Daughters exert a tremendous influence on the progressing social and economic orders. They are potentially immortal, at least until such time as they elect to humanize, receive Adjusters, and start for Paradise.

On the evolutionary worlds a being must humanize to receive a Thought Adjuster. All ascendant members of the Mortal Corps of Finaliters have been Adjuster indwelt and fused except seraphim, and they are Father indwelt by another type of spirit at the time of being mustered into this corps.

5. THE ACME OF MATERIAL DEVELOPMENT

Mortal creatures living on a sin-stricken, evil-dominated, self-seeking, isolated world, such as Urantia, can hardly conceive of the physical perfection, the intellectual attainment, and the spiritual development which characterize these advanced epochs of evolution on a sinless sphere.

The advanced stages of a world settled in light and life represent the acme of evolutionary material development. On these cultured worlds, gone are the idleness and friction of the earlier primitive ages. Poverty and social inequality have all but vanished, degeneracy has disappeared, and delinquency is rarely observed. Insanity has practically ceased to exist, and feeble-mindedness is a rarity.

The economic, social, and administrative status of these worlds is of a high and perfected order. Science, art, and industry flourish, and society is a smoothly working mechanism of high material, intellectual, and cultural achievement. Industry has been largely diverted to serving the higher aims of such a superb civilization. The economic life of such a world has become ethical.

War has become a matter of history, and there are no more armies or police forces. Government is gradually disappearing. Self-control is slowly rendering laws of human enactment obsolete. The extent of civil government and statutory regulation, in an intermediate state of advancing civilization, is in inverse proportion to the morality and spirituality of the citizenship.

Schools are vastly improved and are devoted to the training of mind and the expansion of soul. The art centers are exquisite and the musical organizations superb. The temples of worship with their associated schools of philosophy and experiential religion are creations of beauty and grandeur. The open-air arenas of worship assembly are equally sublime in the simplicity of their artistic appointment.

The provisions for competitive play, humor, and other phases of personal and group achievement are ample and appropriate. A special feature of the competitive activities on such a highly cultured world concerns the efforts of individuals and groups to excel in the sciences and philosophies of cosmology. Literature and oratory flourish, and language is so improved as to be symbolic of concepts as well as to be expressive of ideas. Life is refreshingly simple; man has at last co-ordinated a high state of mechanical development with an inspiring intellectual attainment and has overshadowed both with an exquisite spiritual achievement. The pursuit of happiness is an experience of joy and satisfaction.

6. The Individual Mortal

As worlds advance in the settled status of light and life, society becomes increasingly peaceful. The individual, while no less independent and devoted to his family, has become more altruistic and fraternal.

On Urantia, and as you are, you can have little appreciation of the advanced status and progressive nature of the enlightened races of these perfected worlds. These people are the flowering of the evolutionary races. But such beings are still mortal; they continue to breathe, eat, sleep, and drink. This great evolution is not heaven, but it is a sublime foreshadowing of the divine worlds of the Paradise ascent.

On a normal world the biologic fitness of the mortal race was long since brought up to a high level during the post-Adamic epochs; and now, from age to age throughout the settled eras the physical evolution of man continues. Both vision and hearing are extended. By now the population has become stationary in numbers. Reproduction is regulated in accordance with planetary requirements and innate hereditary endowments: The mortals on a planet during this age are divided into from five to ten groups, and the lower groups are permitted to produce only one half as many children as the higher. The continued improvement of such a magnificent race throughout the era of light and life is largely a matter of the selective reproduction of those racial strains which exhibit superior qualities of a social, philosophic, cosmic, and spiritual nature.

The Adjusters continue to come as in former evolutionary eras, and as the epochs pass, these mortals are increasingly able to commune with the indwelling Father fragment. During the embryonic and prespiritual stages of development the adjutant mind-spirits are still functioning. The Holy Spirit and the ministry of angels are even more effective as the successive epochs of settled life are experienced. In the fourth stage of light and life the advanced mortals seem to experience considerable conscious contact with the spirit presence of the Master Spirit of superuniverse jurisdiction, while the philosophy of such a world is focused upon the attempt to comprehend the new revelations of God the Supreme. More than one half of the human inhabitants on planets of this advanced status experience translation to the morontia state from among the living. Even so, "old things are passing away; behold, all things are becoming new."

We conceive that physical evolution will have attained its full development by the end of the fifth epoch of the light-and-life era. We observe that the upper limits of spiritual development associated with evolving human mind are determined by the Adjuster-fusion level of conjoint morontia values and cosmic meanings. But concerning wisdom: While we do not really know, we conjecture that there can never be a limit to intellectual evolution and the attainment of wisdom. On a seventh-stage world, wisdom can exhaust the material potentials, enter upon mota insight, and eventually even taste of absonite grandeur.

We observe that on these highly evolved and long seventh-stage worlds human beings fully learn the local universe language before they are translated; and I have visited a few very old planets where abandonters were teaching the older mortals the tongue of the superuniverse. And on these worlds I have observed the technique whereby the absonite personalities reveal the presence of the finaliters in the morontia temple.

This is the story of the magnificent goal of mortal striving on the evolutionary worlds; and it all takes place even before human beings enter upon their morontia careers; all of this splendid development is attainable by material mortals on the inhabited worlds, the very first stage of that endless and incomprehensible career of Paradise ascension and divinity attainment.

But can you possibly imagine what sort of evolutionary mortals are now coming up from worlds long existing in the seventh epoch of settled light and life? It is such as these who go on to the morontia worlds of the local universe capital to begin their ascension careers.

If the mortals of distraught Urantia could only view one of these more advanced worlds long settled in light and life, they would nevermore question the wisdom of the evolutionary scheme of creation. Were there no future of eternal creature progression, still the superb evolutionary attainments of the mortal races on such settled worlds of perfected achievement would amply justify man's creation on the worlds of time and space.

We often ponder: If the grand universe should be settled in light and life, would the ascending exquisite mortals still be destined to the Corps of the Finality? But we do not know.

7. THE FIRST OR PLANETARY STAGE

This epoch extends from the appearance of the morontia temple at the new planetary headquarters to the time of the settling of the entire system in light and life. This age is inaugurated by the Trinity

Teacher Sons at the close of their successive world missions when the Planetary Prince is elevated to the status of Planetary Sovereign by the mandate and personal presence of the Paradise bestowal Son of that sphere. Concomitant therewith the finaliters inaugurate their active participation in planetary affairs.

To outward and visible appearances the actual rulers, or directors, of such a world settled in light and life are the Material Son and Daughter, the Planetary Adam and Eve. The finaliters are invisible, as also is the Prince-Sovereign except when in the morontia temple. The actual and literal heads of the planetary regime are therefore the Material Son and Daughter. It is the knowledge of these arrangements that has given prestige to the idea of kings and queens throughout the universe realms. And kings and queens are a great success under these ideal circumstances, when a world can command such high personalities to act in behalf of still higher but invisible rulers.

When such an era is attained on your world, no doubt Machiventa Melchizedek, now the vicegerent Planetary Prince of Urantia, will occupy the seat of the Planetary Sovereign; and it has long been conjectured on Jerusem that he will be accompanied by a son and daughter of the Urantia Adam and Eve who are now held on Edentia as wards of the Most Highs of Norlatiadek. These children of Adam might so serve on Urantia in association with the Melchizedek-Sovereign since they were deprived of procreative powers almost 37,000 years ago at the time they gave up their material bodies on Urantia in preparation for transit to Edentia.

This settled age continues on and on until every inhabited planet in the system attains the era of stabilization; and then, when the youngest world—the last to achieve light and life—has experienced such settledness for one millennium of system time, the entire system enters the stabilized status, and the individual worlds are ushered into the system epoch of the era of light and life.

8. THE SECOND OR SYSTEM STAGE

When an entire system becomes settled in life, a new order of government is inaugurated. The Planetary Sovereigns become members of the system conclave, and this new administrative body, subject to the veto of the Constellation Fathers, is supreme in authority. Such a system of inhabited worlds becomes virtually self-governing. The system legislative assembly is constituted on the headquarters world, and each planet sends its ten representatives thereto. Courts are now established on the system capitals, and only appeals are taken to the universe headquarters.

With the settling of the system the Assigned Sentinel, representative of the superuniverse Supreme Executive, becomes the volunteer adviser to the system supreme court and actual presiding officer of the new legislative assembly.

After the settling of an entire system in light and life the System Sovereigns will no more come and go. Such a sovereign remains perpetually at the head of his system. The assistant sovereigns continue to change as in former ages.

During this epoch of stabilization, for the first time midsoniters come from the universe headquarters worlds of their sojourn to act as counselors to the legislative assemblies and advisers to the adjudicational tribunals. These midsoniters also carry on certain efforts to inculcate new mota meanings of supreme value into the teaching enterprises which they sponsor jointly with the finaliters. What the Material Sons did for the mortal races biologically, the midsonite creatures now do for these unified and glorified humans in the ever-advancing realms of philosophy and spiritualized thinking.

On the inhabited worlds the Teacher Sons become voluntary collaborators with the finaliters, and these same Teacher Sons also accompany the finaliters to the mansion worlds when those spheres are no longer to be utilized as differential receiving worlds after an entire system is settled in light and life; at least this is true by the time the entire constellation has thus evolved. But there are no groups that far advanced in Nebadon.

We are not permitted to reveal the nature of the work of the finaliters who will supervise such rededicated mansion worlds. You have, however, been informed that there are throughout the universes various types of intelligent creatures who have not been portrayed in these narratives.

And now, as the systems one by one become settled in light by virtue of the progress of their component worlds, the time comes when the last system in a given constellation attains stabilization, and the universe administrators—the Master Son, the Union of Days, and the Bright and Morning Star—arrive on the capital of the constellation to proclaim the Most Highs the unqualified rulers of the newly perfected family of one hundred settled systems of inhabited worlds.

9. THE THIRD OR CONSTELLATION STAGE

The unification of a whole constellation of settled systems is attended by new distributions of executive authority and additional readjustments of universe administration. This epoch witnesses advanced attainment on every inhabited world but is particularly characterized by readjustments on the constellation headquarters, with marked modification of relationships with both the system supervision and the local universe government. During this age many constellation and universe activities are transferred to the system capitals, and the representatives of the superuniverse assume new and more intimate relations with the planetary, system, and universe rulers. Concomitant with these new associations, certain superuniverse administrators establish themselves on the constellation capitals as volunteer advisers to the Most High Fathers.

When a constellation is thus settled in light, the legislative function ceases, and the house of System Sovereigns, presided over by the Most Highs, functions instead. Now, for the first time, such administrative groups deal directly with the superuniverse government in matters pertaining to Havona and Paradise relationships. Otherwise the constellation remains related to the local universe as before. From stage to stage in the settled life the univitatia continue to administer the constellation morontia worlds.

As the ages pass, the Constellation Fathers take over more and more of the detailed administrative or supervising functions which were formerly centered on the universe headquarters. By the attainment of the sixth stage of stabilization these unified constellations will have reached the position of well-nigh complete autonomy. Entrance upon the seventh stage of settledness will no doubt witness the exaltation of these rulers to the true dignity signified by their names, the Most Highs. To all intents and purposes the constellations will then deal directly with the superuniverse rulers, while the local universe government will expand to grasp the responsibilities of new grand universe obligations.

10. THE FOURTH OR LOCAL UNIVERSE STAGE

When a universe becomes settled in light and life, it soon swings into the established superuniverse circuits, and the Ancients of Days proclaim the establishment of *the supreme council of unlimited authority*. This new governing body consists of the one hundred Faithfuls of Days, presided over by the Union of Days, and the first act of this supreme council is to acknowledge the continued sovereignty of the Master Creator Son.

The universe administration, as far as concerns Gabriel and the Father Melchizedek, is quite unchanged. This council of unlimited authority is chiefly concerned with the new problems and the new conditions arising out of the advanced status of light and life.

The Associate Inspector now mobilizes all Assigned Sentinels to constitute *the stabilization corps of the local universe* and asks the Father Melchizedek to share its supervision with him. And now, for the first time, a corps of the Inspired Trinity Spirits are assigned to the service of the Union of Days.

The settling of an entire local universe in light and life inaugurates profound readjustments in the entire scheme of administration, from the individual inhabited worlds to the universe headquarters. New relationships extend down to the constellations and systems. The local universe Mother Spirit experiences new liaison relations with the Master Spirit of the superuniverse, and Gabriel establishes direct contact with the Ancients of Days to be effective when and as the Master Son may be absent from the headquarters world.

During this and subsequent ages the Magisterial Sons continue to function as dispensational adjudicators, while one hundred of these Avonal Sons of Paradise constitute the new high council of the Bright and Morning Star on the universe capital. Later on, and as requested by the System Sovereigns, one of these Magisterial Sons will become the supreme counselor stationed on the headquarters world of each local system until the seventh stage of unity is attained.

During this epoch the Trinity Teacher Sons are volunteer advisers, not only to the Planetary Sovereigns, but in groups of three they similarly serve the Constellation Fathers. And at last these Sons find their place in the local universe, for at this time they are removed from the jurisdiction of the local creation and are assigned to the service of the supreme council of unlimited authority.

The finaliter corps now, for the first time, acknowledges the jurisdiction of an extra-Paradise authority, the supreme council. Heretofore the finaliters have recognized no supervision this side of Paradise.

The Creator Sons of such settled universes spend much of their time on Paradise and its associated worlds and in counseling the numerous finaliter groups serving throughout the local creation. In this way the man of Michael will find a fuller fraternity of association with the glorified finaliter mortals.

Speculation concerning the function of these Creator Sons in connection with the outer universes now in process of preliminary assembly is wholly futile. But we all engage in such postulations from time to time. On attaining this fourth stage of development the Creator Son becomes administratively free; the Divine Minister is progressively blending her ministry with that of the superuniverse Master Spirit and the Infinite Spirit. There seems to be evolving a new and sublime relationship between the Creator Son, the Creative Spirit, the Evening Stars, the Teacher Sons, and the ever-increasing finaliter corps.

If Michael should ever leave Nebadon, Gabriel would undoubtedly become chief administrator with the Father Melchizedek as his associate. At the same time new status would be imparted to all orders of permanent citizenship, such as Material Sons, univitatia, midsoniters, susatia, and Spirit-fused mortals. But as long as evolution continues, the seraphim and the archangels will be required in universe administration.

We are, however, satisfied regarding two features of our speculations: If the Creator Sons are destined to the outer universes, the Divine Ministers will undoubtedly accompany them. We are equally sure that the Melchizedeks are to remain with the universes of their origin. We hold that the Melchizedeks are destined to play ever-increasingly responsible parts in local universe government and administration.

11. THE MINOR AND MAJOR SECTOR STAGES

Minor and major sectors of the superuniverse do not figure directly in the plan of being settled in light and life. Such an evolutionary progression pertains primarily to the local universe as a unit and concerns only the components of a local universe. A superuniverse is settled in light and life when all of its component local universes are thus perfected. But not one of the seven superuniverses has attained a level of progression even approaching this.

The minor sector age. As far as observations can penetrate, the fifth or minor sector stage of stabilization has exclusively to do with physical status and with the co-ordinate settling of the one hundred associated local universes in the established circuits of the superuniverse. Apparently none but the power centers and their associates are concerned in these realignments of the material creation.

The major sector age. Concerning the sixth stage, or major sector stabilization, we can only conjecture since none of us have witnessed such an event. Nevertheless, we can postulate much concerning the administrative and other readjustments which would probably accompany such an advanced status of inhabited worlds and their universe groupings.

Since the minor sector status has to do with co-ordinate physical equilibrium, we infer that major sector unification will be concerned with certain new intellectual levels of attainment, possibly some advanced achievements in the supreme realization of cosmic wisdom.

We arrive at conclusions regarding the readjustments which would probably attend the realization of hitherto unattained levels of evolutionary progress by observing the results of such achievements on the individual worlds and in the experiences of individual mortals living on these older and highly developed spheres.

Let it be made clear that the administrative mechanisms and governmental techniques of a universe or a superuniverse cannot in any manner limit or retard the evolutionary development or spiritual progress of an individual inhabited planet or of any individual mortal on such a sphere.

In some of the older universes we find worlds settled in the fifth and the sixth stages of light and life—even far extended into the seventh epoch—whose local systems are not yet settled in light. Younger planets may delay system unification, but this does not in the least handicap the progress of an older and advanced world. Neither can environmental limitations, even on an isolated world, thwart the personal attainment of the individual mortal; Jesus of Nazareth, as a man among men, personally achieved the status of light and life over nineteen hundred years ago on Urantia.

It is by observing what takes place on long-settled worlds that we arrive at fairly reliable conclusions as to what will happen when a whole superuniverse is settled in light, even if we cannot safely postulate the event of the stabilization of the seven superuniverses.

12. The Seventh or Superuniverse Stage

We cannot positively forecast what would occur when a superuniverse became settled in light because such an event has never factualized. From the teachings of the Melchizedeks, which have never been contradicted, we infer that sweeping changes would be made in the entire organization and administration of every unit of the creations of time and space extending from the inhabited worlds to the superuniverse headquarters.

It is generally believed that large numbers of the otherwise unattached creature-trinitized sons are to be assembled on the headquarters and divisional capitals of the settled superuniverses. This may be in anticipation of the sometime arrival of outer-spacers on their way in to Havona and Paradise; but we really do not know.

If and when a superuniverse should be settled in light and life, we believe that the now advisory Unqualified Supervisors of the Supreme would become the high administrative body on the headquarters world of the superuniverse. These are the personalities who are able to contact directly with the absonite administrators, who will forthwith become active in the settled superuniverse. Although these Unqualified Supervisors have long functioned as advisers and counselors in advanced evolutionary units of creation, they do not assume administrative responsibilities until the authority of the Supreme Being becomes sovereign.

The Unqualified Supervisors of the Supreme, who function more extensively during this epoch, are not finite, absonite, ultimate, or infinite; they *are* supremacy and only represent God the Supreme. They are the personalization of time-space supremacy and therefore do not function in Havona. They function only as supreme unifiers. They may possibly be involved in the technique of universe reflectivity, but we are not certain.

None of us entertain a satisfactory concept of what will happen when the grand universe (the seven superuniverses as dependent on Havona) becomes entirely settled in light and life. That event will undoubtedly be the most profound occurrence in the annals of eternity since the appearance of the central universe. There are those who hold that the Supreme Being himself will emerge from the Havona mystery enshrouding his spirit person and will become residential on the headquarters of the seventh superuniverse as the almighty and experiential sovereign of the perfected creations of time and space. But we really do not know.

[Presented by a Mighty Messenger temporarily assigned to the Archangel Council on Urantia.]

SIXTEEN

GOVERNMENT ON
A NEIGHBORING PLANET

(PAPER 72)

BY PERMISSION of Lanaforge and with the approval of the Most
Highs of Edentia, I am authorized to narrate something of the
social, moral, and political life of the most advanced human race living on a not far-distant planet belonging to the Satania system.

Of all the Satania worlds which became isolated because of
participation in the Lucifer rebellion, this planet has experienced a
history most like that of Urantia. The similarity of the two spheres
undoubtedly explains why permission to make this extraordinary
presentation was granted, for it is most unusual for the system rulers
to consent to the narration on one planet of the affairs of another.

This planet, like Urantia, was led astray by the disloyalty of its
Planetary Prince in connection with the Lucifer rebellion. It received
a Material Son shortly after Adam came to Urantia, and this Son also
defaulted, leaving the sphere isolated, since a Magisterial Son has
never been bestowed upon its mortal races.

1. THE CONTINENTAL NATION

Notwithstanding all these planetary handicaps a very superior
civilization is evolving on an isolated continent about the size of
Australia. This nation numbers about 140 million. Its people are a
mixed race, predominantly blue and yellow, having a slightly greater

proportion of violet than the so-called white race of Urantia. These different races are not yet fully blended, but they fraternize and socialize very acceptably. The average length of life on this continent is now ninety years, fifteen per cent higher than that of any other people on the planet.

The industrial mechanism of this nation enjoys a certain great advantage derived from the unique topography of the continent. The high mountains, on which heavy rains fall eight months in the year, are situated at the very center of the country. This natural arrangement favors the utilization of water power and greatly facilitates the irrigation of the more arid western quarter of the continent.

These people are self-sustaining, that is, they can live indefinitely without importing anything from the surrounding nations. Their natural resources are replete, and by scientific techniques they have learned how to compensate for their deficiencies in the essentials of life. They enjoy a brisk domestic commerce but have little foreign trade owing to the universal hostility of their less progressive neighbors.

This continental nation, in general, followed the evolutionary trend of the planet: The development from the tribal stage to the appearance of strong rulers and kings occupied thousands of years. The unconditional monarchs were succeeded by many different orders of government—abortive republics, communal states, and dictators came and went in endless profusion. This growth continued until about five hundred years ago when, during a politically fermenting period, one of the nation's powerful dictator-triumvirs had a change of heart. He volunteered to abdicate upon condition that one of the other rulers, the baser of the remaining two, also vacate his dictatorship. Thus was the sovereignty of the continent placed in the hands of one ruler. The unified state progressed under strong monarchial rule for over one hundred years, during which there evolved a masterful charter of liberty.

The subsequent transition from monarchy to a representative form of government was gradual, the kings remaining as mere social or sentimental figureheads, finally disappearing when the male line of descent ran out. The present republic has now been in existence just two hundred years, during which time there has been a continuous progression toward the governmental techniques about to be narrated, the last developments in industrial and political realms having been made within the past decade.

2. Political Organization

This continental nation now has a representative government with a centrally located national capital. The central government consists of a strong federation of one hundred comparatively free states. These states elect their governors and legislators for ten years, and none are eligible for re-election. State judges are appointed for life by the governors and confirmed by their legislatures, which consist of one representative for each one hundred thousand citizens.

There are five different types of metropolitan government, depending on the size of the city, but no city is permitted to have more than one million inhabitants. On the whole, these municipal governing schemes are very simple, direct, and economical. The few offices of city administration are keenly sought by the highest types of citizens.

The federal government embraces three co-ordinate divisions: executive, legislative, and judicial. The federal chief executive is elected every six years by universal territorial suffrage. He is not eligible for re-election except upon the petition of at least seventy-five state legislatures concurred in by the respective state governors, and then but for one term. He is advised by a supercabinet composed of all living ex-chief executives.

The legislative division embraces three houses:

1. *The upper house* is elected by industrial, professional, agricultural, and other groups of workers, balloting in accordance with economic function.

2. *The lower house* is elected by certain organizations of society embracing the social, political, and philosophic groups not included in industry or the professions. All citizens in good standing participate in the election of both classes of representatives, but they are differently grouped, depending on whether the election pertains to the upper or lower house.

3. *The third house*—the elder statesmen—embraces the veterans of civic service and includes many distinguished persons nominated by the chief executive, by the regional (subfederal) ex-

ecutives, by the chief of the supreme tribunal, and by the presiding officers of either of the other legislative houses. This group is limited to one hundred, and its members are elected by the majority action of the elder statesmen themselves. Membership is for life, and when vacancies occur, the person receiving the largest ballot among the list of nominees is thereby duly elected. The scope of this body is purely advisory, but it is a mighty regulator of public opinion and exerts a powerful influence upon all branches of the government.

Very much of the federal administrative work is carried on by the ten regional (subfederal) authorities, each consisting of the association of ten states. These regional divisions are wholly executive and administrative, having neither legislative nor judicial functions. The ten regional executives are the personal appointees of the federal chief executive, and their term of office is concurrent with his—six years. The federal supreme tribunal approves the appointment of these ten regional executives, and while they may not be reappointed, the retiring executive automatically becomes the associate and adviser of his successor. Otherwise, these regional chiefs choose their own cabinets of administrative officials.

This nation is adjudicated by two major court systems—the law courts and the socioeconomic courts. The law courts function on the following three levels:

1. *Minor courts* of municipal and local jurisdiction, whose decisions may be appealed to the high state tribunals.

2. *State supreme courts*, whose decisions are final in all matters not involving the federal government or jeopardy of citizenship rights and liberties. The regional executives are empowered to bring any case at once to the bar of the federal supreme court.

3. *Federal supreme court*—the high tribunal for the adjudication of national contentions and the appellate cases coming up from the state courts. This supreme tribunal consists of twelve men over forty and under seventy-five years of age who have served two or more years on some state tribunal, and who have been appointed to this high position by the chief executive with the majority approval of the supercabinet and the third house of the legislative assembly. All decisions of this supreme judicial body are by at least a two-thirds vote.

The socioeconomic courts function in the following three divisions:

1. *Parental courts,* associated with the legislative and executive divisions of the home and social system.

2. *Educational courts*—the juridical bodies connected with the state and regional school systems and associated with the executive and legislative branches of the educational administrative mechanism.

3. *Industrial courts*—the jurisdictional tribunals vested with full authority for the settlement of all economic misunderstandings.

The federal supreme court does not pass upon socioeconomic cases except upon the three-quarters vote of the third legislative branch of the national government, the house of elder statesmen. Otherwise, all decisions of the parental, educational, and industrial high courts are final.

3. THE HOME LIFE

On this continent it is against the law for two families to live under the same roof. And since group dwellings have been outlawed, most of the tenement type of buildings have been demolished. But the unmarried still live in clubs, hotels, and other group dwellings. The smallest homesite permitted must provide fifty thousand square feet of land. All land and other property used for home purposes are free from taxation up to ten times the minimum homesite allotment.

The home life of this people has greatly improved during the last century. Attendance of parents, both fathers and mothers, at the parental schools of child culture is compulsory. Even the agriculturists who reside in small country settlements carry on this work by correspondence, going to the near-by centers for oral instruction once in ten days—every two weeks, for they maintain a five-day week.

The average number of children in each family is five, and they are under the full control of their parents or, in case of the demise of one or both, under that of the guardians designated by the parental courts. It is considered a great honor for any family to be awarded the guardianship of a full orphan. Competitive examinations are held among parents, and the orphan is awarded to the home of those displaying the best parental qualifications.

These people regard the home as the basic institution of their civilization. It is expected that the most valuable part of a child's education and character training will be secured from his parents and at home, and fathers devote almost as much attention to child culture as do mothers.

All sex instruction is administered in the home by parents or by legal guardians. Moral instruction is offered by teachers during the rest periods in the school shops, but not so with religious training, which is deemed to be the exclusive privilege of parents, religion being looked upon as an integral part of home life. Purely religious instruction is given publicly only in the temples of philosophy, no such exclusively religious institutions as the Urantia churches having developed among this people. In their philosophy, religion is the striving to know God and to manifest love for one's fellows through service for them, but this is not typical of the religious status of the other nations on this planet. Religion is so entirely a family matter among these people that there are no public places devoted exclusively to religious assembly. Politically, church and state, as Urantians are wont to say, are entirely separate, but there is a strange overlapping of religion and philosophy.

Until twenty years ago the spiritual teachers (comparable to Urantia pastors), who visit each family periodically to examine the children to ascertain if they have been properly instructed by their parents, were under governmental supervision. These spiritual advisers and examiners are now under the direction of the newly created Foundation of Spiritual Progress, an institution supported by voluntary contributions. Possibly this institution may not further evolve until after the arrival of a Paradise Magisterial Son.

Children remain legally subject to their parents until they are fifteen, when the first initiation into civic responsibility is held. Thereafter, every five years for five successive periods similar public exercises are held for such age groups at which their obligations to parents are lessened, while new civic and social responsibilities to the state are assumed. Suffrage is conferred at twenty, the right to marry without parental consent is not bestowed until twenty-five, and children must leave home on reaching the age of thirty.

Marriage and divorce laws are uniform throughout the nation. Marriage before twenty—the age of civil enfranchisement—is not permitted. Permission to marry is only granted after one year's notice of intention, and after both bride and groom present certificates showing that they have been duly instructed in the parental schools regarding the responsibilities of married life.

Divorce regulations are somewhat lax, but decrees of separation, issued by the parental courts, may not be had until one year after application therefor has been recorded, and the year on this planet is considerably longer than on Urantia. Notwithstanding their easy divorce laws, the present rate of divorces is only one tenth that of the civilized races of Urantia.

4. THE EDUCATIONAL SYSTEM

The educational system of this nation is compulsory and coeducational in the precollege schools that the student attends from the ages of five to eighteen. These schools are vastly different from those of Urantia. There are no classrooms, only one study is pursued at a time, and after the first three years all pupils become assistant teachers, instructing those below them. Books are used only to secure information that will assist in solving the problems arising in the school shops and on the school farms. Much of the furniture used on the continent and the many mechanical contrivances—this is a great age of invention and mechanization—are produced in these shops. Adjacent to each shop is a working library where the student may consult the necessary reference books. Agriculture and horticulture are also taught throughout the entire educational period on the extensive farms adjoining every local school.

The feeble-minded are trained only in agriculture and animal husbandry, and are committed for life to special custodial colonies where they are segregated by sex to prevent parenthood, which is denied all subnormals. These restrictive measures have been in operation for seventy-five years; the commitment decrees are handed down by the parental courts.

Everyone takes one month's vacation each year. The precollege schools are conducted for nine months out of the year of ten, the vacation being spent with parents or friends in travel. This travel is a part of the adult-education program and is continued throughout a lifetime, the funds for meeting such expenses being accumulated by the same methods as those employed in old-age insurance.

One quarter of the school time is devoted to play—competitive athletics—the pupils progressing in these contests from the local, through the state and regional, and on to the national trials of skill and prowess. Likewise, the oratorical and musical contests, as well as those in science and philosophy, occupy the attention of students from the lower social divisions on up to the contests for national honors.

The school government is a replica of the national government with its three correlated branches, the teaching staff functioning as the third or advisory legislative division. The chief object of education on this continent is to make every pupil a self-supporting citizen.

Every child graduating from the precollege school system at eighteen is a skilled artisan. Then begins the study of books and the pursuit of special knowledge, either in the adult schools or in the colleges. When a brilliant student completes his work ahead of schedule, he is granted an award of time and means wherewith he may execute some pet project of his own devising. The entire educational system is designed to adequately train the individual.

5. Industrial Organization

The industrial situation among this people is far from their ideals; capital and labor still have their troubles, but both are becoming adjusted to the plan of sincere co-operation. On this unique continent the workers are increasingly becoming shareholders in all industrial concerns; every intelligent laborer is slowly becoming a small capitalist.

Social antagonisms are lessening, and good will is growing apace. No grave economic problems have arisen out of the abolition of slavery (over one hundred years ago) since this adjustment was effected gradually by the liberation of two per cent each year. Those slaves who satisfactorily passed mental, moral, and physical tests were granted citizenship; many of these superior slaves were war captives or children of such captives. Some fifty years ago they deported the last of their inferior slaves, and still more recently they are addressing themselves to the task of reducing the numbers of their degenerate and vicious classes.

These people have recently developed new techniques for the adjustment of industrial misunderstandings and for the correction of economic abuses which are marked improvements over their older methods of settling such problems. Violence has been outlawed as a procedure in adjusting either personal or industrial differences. Wages, profits, and other economic problems are not rigidly regulated, but they are in general controlled by the industrial legislatures, while all disputes arising out of industry are passed upon by the industrial courts.

The industrial courts are only thirty years old but are functioning very satisfactorily. The most recent development provides that hereafter the industrial courts shall recognize legal compensation as falling in three divisions:

1. Legal rates of interest on invested capital.

2. Reasonable salary for skill employed in industrial operations.

3. Fair and equitable wages for labor.

These shall first be met in accordance with contract, or in the face of decreased earnings they shall share proportionally in transient reduction. And thereafter all earnings in excess of these fixed charges shall be regarded as dividends and shall be prorated to all three divisions: capital, skill, and labor.

Every ten years the regional executives adjust and decree the lawful hours of daily gainful toil. Industry now operates on a five-day week, working four and playing one. These people labor six hours each working day and, like students, nine months in the year of ten. Vacation is usually spent in travel, and new methods of transportation having been so recently developed, the whole nation is travel bent. The climate favors travel about eight months in the year, and they are making the most of their opportunities.

Two hundred years ago the profit motive was wholly dominant in industry, but today it is being rapidly displaced by other and higher driving forces. Competition is keen on this continent, but much of it has been transferred from industry to play, skill, scientific achievement, and intellectual attainment. It is most active in social service and governmental loyalty. Among this people public service is rapidly becoming the chief goal of ambition. The richest man on the continent works six hours a day in the office of his machine shop and then hastens over to the local branch of the school of statesmanship, where he seeks to qualify for public service.

Labor is becoming more honorable on this continent, and all able-bodied citizens over eighteen work either at home and on farms, at some recognized industry, on the public works where the temporarily unemployed are absorbed, or else in the corps of compulsory laborers in the mines.

These people are also beginning to foster a new form of social disgust—disgust for both idleness and unearned wealth. Slowly but certainly they are conquering their machines. Once they, too, struggled for political liberty and subsequently for economic freedom. Now are they entering upon the enjoyment of both while in addition they are beginning to appreciate their well-earned leisure, which can be devoted to increased self-realization.

6. Old-Age Insurance

This nation is making a determined effort to replace the self-respect-destroying type of charity by dignified government-insurance guarantees of security in old age. This nation provides every child an education and every man a job; therefore can it successfully carry out such an insurance scheme for the protection of the infirm and aged.

Among this people all persons must retire from gainful pursuit at sixty-five unless they secure a permit from the state labor commissioner which will entitle them to remain at work until the age of seventy. This age limit does not apply to government servants or philosophers. The physically disabled or permanently crippled can be placed on the retired list at any age by court order countersigned by the pension commissioner of the regional government.

The funds for old-age pensions are derived from four sources:

1. One day's earnings each month are requisitioned by the federal government for this purpose, and in this country everybody works.

2. Bequests—many wealthy citizens leave funds for this purpose.

3. The earnings of compulsory labor in the state mines. After the conscript workers support themselves and set aside their own retirement contributions, all excess profits on their labor are turned over to this pension fund.

4. The income from natural resources. All natural wealth on the continent is held as a social trust by the federal government, and the income therefrom is utilized for social purposes, such as disease prevention, education of geniuses, and expenses of especially promising individuals in the statesmanship schools. One half of the income from natural resources goes to the old-age pension fund.

Although state and regional actuarial foundations supply many forms of protective insurance, old-age pensions are solely administered by the federal government through the ten regional departments.

These government funds have long been honestly administered. Next to treason and murder, the heaviest penalties meted out by the courts are attached to betrayal of public trust. Social and political disloyalty are now looked upon as being the most heinous of all crimes.

7. Taxation

The federal government is paternalistic only in the administration of old-age pensions and in the fostering of genius and creative originality; the state governments are slightly more concerned with the individual citizen, while the local governments are much more paternalistic or socialistic. The city (or some subdivision thereof) concerns itself with such matters as health, sanitation, building regulations, beautification, water supply, lighting, heating, recreation, music, and communication.

In all industry first attention is paid to health; certain phases of physical well-being are regarded as industrial and community prerogatives, but individual and family health problems are matters of personal concern only. In medicine, as in all other purely personal matters, it is increasingly the plan of government to refrain from interfering.

Cities have no taxing power, neither can they go in debt. They receive per capita allowances from the state treasury and must supplement such revenue from the earnings of their socialistic enterprises and by licensing various commercial activities.

The rapid-transit facilities, which make it practical greatly to extend the city boundaries, are under municipal control. The city fire departments are supported by the fire-prevention and insurance foundations, and all buildings, in city or country, are fireproof—have been for over seventy-five years.

There are no municipally appointed peace officers; the police forces are maintained by the state governments. This department is recruited almost entirely from the unmarried men between twenty-five and fifty. Most of the states assess a rather heavy bachelor tax, which is remitted to all men joining the state police. In the average state the police force is now only one tenth as large as it was fifty years ago.

There is little or no uniformity among the taxation schemes of the one hundred comparatively free and sovereign states as economic and other conditions vary greatly in different sections of the continent. Every state has ten basic constitutional provisions which cannot be modified except by consent of the federal supreme court, and one of these articles prevents levying a tax of more than one per cent on the value of any property in any one year, homesites, whether in city or country, being exempted.

The federal government cannot go in debt, and a three-fourths referendum is required before any state can borrow except for purposes of war. Since the federal government cannot incur debt, in the event of war the National Council of Defense is empowered to assess the states for money, as well as for men and materials, as it may be required. But no debt may run for more than twenty-five years.

Income to support the federal government is derived from the following five sources:

1. *Import duties.* All imports are subject to a tariff designed to protect the standard of living on this continent, which is far above that of any other nation on the planet. These tariffs are set by the highest industrial court after both houses of the industrial congress have ratified the recommendations of the chief executive of economic affairs, who is the joint appointee of these two legislative bodies. The upper industrial house is elected by labor, the lower by capital.

2. *Royalties.* The federal government encourages invention and original creations in the ten regional laboratories, assisting all types of geniuses—artists, authors, and scientists—and protecting their patents. In return the government takes one half the profits

realized from all such inventions and creations, whether pertaining to machines, books, artistry, plants, or animals.

3. *Inheritance tax.* The federal government levies a graduated inheritance tax ranging from one to fifty per cent, depending on the size of an estate as well as on other conditions.

4. *Military equipment.* The government earns a considerable sum from the leasing of military and naval equipment for commercial and recreational usages.

5. *Natural resources.* The income from natural resources, when not fully required for the specific purposes designated in the charter of federal statehood, is turned into the national treasury.

Federal appropriations, except war funds assessed by the National Council of Defense, are originated in the upper legislative house, concurred in by the lower house, approved by the chief executive, and finally validated by the federal budget commission of one hundred. The members of this commission are nominated by the state governors and elected by the state legislatures to serve for twenty-four years, one quarter being elected every six years. Every six years this body, by a three-fourths ballot, chooses one of its number as chief, and he thereby becomes director-controller of the federal treasury.

8. THE SPECIAL COLLEGES

In addition to the basic compulsory education program extending from the ages of five to eighteen, special schools are maintained as follows:

1. *Statesmanship schools.* These schools are of three classes: national, regional, and state. The public offices of the nation are grouped in four divisions. The first division of public trust pertains principally to the national administration, and all officeholders of this group must be graduates of both regional and national schools of statesmanship. Individuals may accept political, elective, or appointive office in the second division upon graduating from any one of the ten regional schools of statesmanship; their trusts concern responsibilities in the regional administration and the state governments. Division three includes state responsibilities, and such officials are only required to have state degrees of statesmanship.

The fourth and last division of officeholders are not required to hold statesmanship degrees, such offices being wholly appointive. They represent minor positions of assistantship, secretaryships, and technical trusts which are discharged by the various learned professions functioning in governmental administrative capacities.

Judges of the minor and state courts hold degrees from the state schools of statesmanship. Judges of the jurisdictional tribunals of social, educational, and industrial matters hold degrees from the regional schools. Judges of the federal supreme court must hold degrees from all these schools of statesmanship.

2. *Schools of philosophy.* These schools are affiliated with the temples of philosophy and are more or less associated with religion as a public function.

3. *Institutions of science.* These technical schools are co-ordinated with industry rather than with the educational system and are administered under fifteen divisions.

4. *Professional training schools.* These special institutions provide the technical training for the various learned professions, twelve in number.

5. *Military and naval schools.* Near the national headquarters and at the twenty-five coastal military centers are maintained those institutions devoted to the military training of volunteer citizens from eighteen to thirty years of age. Parental consent is required before twenty-five in order to gain entrance to these schools.

9. The Plan of Universal Suffrage

Although candidates for all public offices are restricted to graduates of the state, regional, or federal schools of statesmanship, the progressive leaders of this nation discovered a serious weakness in their plan of universal suffrage and about fifty years ago made constitutional provision for a modified scheme of voting which embraces the following features:

1. Every man and woman of twenty years and over has one vote. Upon attaining this age, all citizens must accept membership in two voting groups: They will join the first in accordance with their economic function—industrial, professional, agricultural, or trade; they will enter the second group according to their political, phil-

osophic, and social inclinations. All workers thus belong to some economic franchise group, and these guilds, like the noneconomic associations, are regulated much as is the national government with its threefold division of powers. Registration in these groups cannot be changed for twelve years.

2. Upon nomination by the state governors or by the regional executives and by the mandate of the regional supreme councils, individuals who have rendered great service to society, or who have demonstrated extraordinary wisdom in government service, may have additional votes conferred upon them not oftener than every five years and not to exceed nine such superfranchises. The maximum suffrage of any multiple voter is ten. Scientists, inventors, teachers, philosophers, and spiritual leaders are also thus recognized and honored with augmented political power. These advanced civic privileges are conferred by the state and regional supreme councils much as degrees are bestowed by the special colleges, and the recipients are proud to attach the symbols of such civic recognition, along with their other degrees, to their lists of personal achievements.

3. All individuals sentenced to compulsory labor in the mines and all governmental servants supported by tax funds are, for the periods of such services, disenfranchised. This does not apply to aged persons who may be retired on pensions at sixty-five.

4. There are five brackets of suffrage reflecting the average yearly taxes paid for each half-decade period. Heavy taxpayers are permitted extra votes up to five. This grant is independent of all other recognition, but in no case can any person cast over ten ballots.

5. At the time this franchise plan was adopted, the territorial method of voting was abandoned in favor of the economic or functional system. All citizens now vote as members of industrial, social, or professional groups, regardless of their residence. Thus the electorate consists of solidified, unified, and intelligent groups who elect only their best members to positions of governmental trust and responsibility. There is one exception to this scheme of functional or group suffrage: The election of a federal chief executive every six years is by nation-wide ballot, and no citizen casts over one vote.

Thus, except in the election of the chief executive, suffrage is exercised by economic, professional, intellectual, and social groupings of the citizenry. The ideal state is organic, and every free and intelligent group of citizens represents a vital and functioning organ within the larger governmental organism.

The schools of statesmanship have power to start proceedings in the state courts looking toward the disenfranchisement of any defective, idle, indifferent, or criminal individual. These people recognize that, when fifty per cent of a nation is inferior or defective and possesses the ballot, such a nation is doomed. They believe the dominance of mediocrity spells the downfall of any nation. Voting is compulsory, heavy fines being assessed against all who fail to cast their ballots.

10. DEALING WITH CRIME

The methods of this people in dealing with crime, insanity, and degeneracy, while in some ways pleasing, will, no doubt, in others prove shocking to most Urantians. Ordinary criminals and the defectives are placed, by sexes, in different agricultural colonies and are more than self-supporting. The more serious habitual criminals and the incurably insane are sentenced to death in the lethal gas chambers by the courts. Numerous crimes aside from murder, including betrayal of governmental trust, also carry the death penalty, and the visitation of justice is sure and swift.

These people are passing out of the negative into the positive era of law. Recently they have gone so far as to attempt the prevention of crime by sentencing those who are believed to be potential murderers and major criminals to life service in the detention colonies. If such convicts subsequently demonstrate that they have become more normal, they may be either paroled or pardoned. The homicide rate on this continent is only one per cent of that among the other nations.

Efforts to prevent the breeding of criminals and defectives were begun over one hundred years ago and have already yielded gratifying results. There are no prisons or hospitals for the insane. For one reason, there are only about ten per cent as many of these groups as are found on Urantia.

11. MILITARY PREPAREDNESS

Graduates of the federal military schools may be commissioned as "guardians of civilization" in seven ranks, in accordance with ability and experience, by the president of the National Council of Defense. This council consists of twenty-five members, nominated by the highest parental, educational, and industrial tribunals, confirmed by the federal supreme court, and presided over ex officio by the chief of staff of co-ordinated military affairs. Such members serve until they are seventy years of age.

The courses pursued by such commissioned officers are four years in length and are invariably correlated with the mastery of some trade or profession. Military training is never given without this associated industrial, scientific, or professional schooling. When military training is finished, the individual has, during his four years' course, received one half of the education imparted in any of the special schools where the courses are likewise four years in length. In this way the creation of a professional military class is avoided by providing this opportunity for a large number of men to support themselves while securing the first half of a technical or professional training.

Military service during peacetime is purely voluntary, and the enlistments in all branches of the service are for four years, during which every man pursues some special line of study in addition to the mastery of military tactics. Training in music is one of the chief pursuits of the central military schools and of the twenty-five training camps distributed about the periphery of the continent. During periods of industrial slackness many thousands of unemployed are automatically utilized in upbuilding the military defenses of the continent on land and sea and in the air.

Although these people maintain a powerful war establishment as a defense against invasion by the surrounding hostile peoples, it may be recorded to their credit that they have not in over one hundred years employed these military resources in an offensive war. They have become civilized to that point where they can vigorously defend civilization without yielding to the temptation to utilize their war powers in aggression. There have been no civil wars since the establishment of the united continental state, but during the last two centuries these people have been called upon to wage nine fierce defensive conflicts, three of which were against mighty confederations of world powers. Although this nation maintains adequate defense against attack by hostile neighbors, it pays far more attention to the training of statesmen, scientists, and philosophers.

When at peace with the world, all mobile defense mechanisms are quite fully employed in trade, commerce, and recreation. When war is declared, the entire nation is mobilized. Throughout the period of hostilities military pay obtains in all industries, and the chiefs of all military departments become members of the chief executive's cabinet.

12. The Other Nations

Although the society and government of this unique people are in many respects superior to those of the Urantia nations, it should be stated that on the other continents (there are eleven on this planet) the governments are decidedly inferior to the more advanced nations of Urantia.

Just now this superior government is planning to establish ambassadorial relations with the inferior peoples, and for the first time a great religious leader has arisen who advocates the sending of missionaries to these surrounding nations. We fear they are about to make the mistake that so many others have made when they have endeavored to force a superior culture and religion upon other races. What a wonderful thing could be done on this world if this continental nation of advanced culture would only go out and bring to itself the best of the neighboring peoples and then, after educating them, send them back as emissaries of culture to their benighted brethren! Of course, if a Magisterial Son should soon come to this advanced nation, great things could quickly happen on this world.

This recital of the affairs of a neighboring planet is made by special permission with the intent of advancing civilization and augmenting governmental evolution on Urantia. Much more could be narrated that would no doubt interest and intrigue Urantians, but this disclosure covers the limits of our permissive mandate.

Urantians should, however, take note that their sister sphere in the Satania family has benefited by neither magisterial nor bestowal missions of the Paradise Sons. Neither are the various peoples of Urantia set off from each other by such disparity of culture as separates the continental nation from its planetary fellows.

The pouring out of the Spirit of Truth provides the spiritual foundation for the realization of great achievements in the interests of the human race of the bestowal world. Urantia is therefore far better prepared for the more immediate realization of a planetary government with its laws, mechanisms, symbols, conventions, and language—all of which could contribute so mightily to the establishment of world-wide peace under law and could lead to the sometime dawning of a real age of spiritual striving; and such an age is the planetary threshold to the utopian ages of light and life.

[Presented by a Melchizedek of Nebadon.]

THE TITLES
OF THE URANTIA PAPERS

PART I—CENTRAL AND SUPERUNIVERSES
1. The Universal Father
2. The Nature of God
3. The Attributes of God
4. God's Relation to the Universe
5. God's Relation to the Individual
6. The Eternal Son
7. Relation of the Eternal Son to the Universe
8. The Infinite Spirit
9. Relation of the Infinite Spirit to the Universe
10. The Paradise Trinity
11. The Eternal Isle of Paradise
12. The Universe of Universes
13. The Sacred Spheres of Paradise
14. The Central and Divine Universe
15. The Seven Superuniverses
16. The Seven Master Spirits
17. The Seven Supreme Spirit Groups
18. The Supreme Trinity Personalities
19. The Co-ordinate Trinity-Origin Beings
20. The Paradise Sons of God
21. The Paradise Creator Sons
22. The Trinitized Sons of God
23. The Solitary Messengers
24. Higher Personalities of the Infinite Spirit
25. The Messenger Hosts of Space

www.ingramcontent.com/pod-product-compliance
Lightning Source LLC
LaVergne TN
LVHW041151080426
835511LV00006B/557